CAMBRIDGE MONOGRAPHS
ON MECHANICS AND APPLIED MATHEMATICS

GENERAL EDITORS

G. K. BATCHELOR, PH.D.
Lecturer in Mathematics in the University of Cambridge

H. BONDI, M.A.
Professor of Applied Mathematics at King's College, University of London

STATIC AND DYNAMIC ELECTRON OPTICS

STATIC AND DYNAMIC ELECTRON OPTICS

AN ACCOUNT OF FOCUSING IN LENS, DEFLECTOR AND ACCELERATOR

BY

P. A. STURROCK, PH.D.

Research Fellow of
St John's College, Cambridge

CAMBRIDGE

AT THE UNIVERSITY PRESS

1955

CAMBRIDGE
UNIVERSITY PRESS

University Printing House, Cambridge CB2 8BS, United Kingdom

Cambridge University Press is part of the University of Cambridge.

It furthers the University's mission by disseminating knowledge in the pursuit of education, learning and research at the highest international levels of excellence.

www.cambridge.org
Information on this title: www.cambridge.org/9781316509784

First published 1955
First paperback edition 2015

A catalogue record for this publication is available from the British Library

ISBN 978-1-316-50978-4 Paperback

CONTENTS

IV The Rotationally Symmetrical System

V Systems of Mirror Symmetry

PART II. DYNAMIC ELECTRON OPTICS

VI Uniform Focusing in Particle Accelerators

VII Periodic Focusing in Particle Accelerators

PREFACE

There have appeared, over the last twenty years, a number of text-books on electron optics, but the great advances which are still going on make it increasingly difficult to include in one volume an adequate account of the theory, design and operation of even the few principal electron-optical instruments. Although there are monographs in English which are confined to design problems or to the use of some specific instrument, the only theoretical monographs so far published have all been written in German.

In writing this book on the theory of electron optics, I have had in mind the needs not only of those already engaged in research and development who wish to calculate the properties of some new instrument or machine, but also of scientists who are not yet familiar with the subject and therefore look to a theoretical exposition for an account of the ideas common to all applications. Part I, on 'static electron optics', deals with the focusing of charged particles in static electromagnetic fields. Part II, on 'dynamic electron optics', is devoted to the focusing of charged particles in time-dependent electromagnetic fields as realized in particle accelerators. The tendency has been, for reasons which I attempt to make clear, to emphasize the optical—rather than the dynamical—aspect of these problems.

Part I began to take shape during the years 1948–51, during which I was carrying out study and research on electron optics at Cambridge, at the National Bureau of Standards, Washington, and at the Laboratoire de Radioélectricité, Paris. My interest in particle accelerators stems from my experience as a research fellow at the Atomic Energy Research Establishment, Harwell, over the years 1951–53.

My aim in writing this book has been three-fold: First, to relate the theory of electron-optical instruments and particle accelerators to classical optics and dynamics by demonstrating the significance and usefulness in these new studies of concepts drawn from the older disciplines, particularly the Lagrange and Poincaré invariants and Hamilton's characteristic functions. Second, to set out general procedures which may be applied to the study of image-formation in electron-optical instruments and stability in particle accelerators; these are based exclusively on the relevant variational principles and make extensive use of modifications of Hamilton's functions which I call 'perturbation characteristic functions'. Third, to apply these procedures to the analysis of the properties of the more important instruments and machines: electron lenses, β-ray and mass spectrographs, cathode-ray-tube deflectors, and certain accelerators including the strong-focusing synchrotron.

The reader need not assume that each chapter is entirely dependent upon preceding chapters. Those without previous knowledge of electron optics may omit Chapter 2 on their first reading, since formulae which are used in Chapters 4 and 5 are derived in Chapter 3. None of Chapter 4 is necessary for the understanding of Chapter 5. If the reader's primary interest is in the dynamics of particle accelerators, he may pass to Part II after reading only the following sections of Part I: 1.1, 1.2, 1.3; 2.1, 2.2, 2.3; 3.1, 3.2, 3.3 and 3.5 as far as equation (3.5.10).

It is a pleasure to thank the many people who have stimulated my interest in electron optics and in particle accelerators, particularly Dr V. E. Cosslett of the Cavendish Laboratory, Cambridge; Dr W. Ehrenberg of Birkbeck College, London; Professor P. Grivet of the Laboratoire de Radioélectricité, Paris; Mr M. E. Haine and Dr G. Liebmann of the A.E.I. Research Laboratory, Aldermaston; Dr L. Marton of the National Bureau of Standards, Washington; and Mr W. Walkinshaw of the Atomic Energy Research Establishment, Harwell. I owe especial thanks to Mr J. S. Bell, also of A.E.R.E., for his detailed criticism of the first draft of Chapters 6 and 7; to Mr A. R. Curtis, now at Sheffield University, for criticism of an early draft of the first three chapters; and to Dr D. Gabor for his extensive criticism of work on which this monograph is in part based. Finally, I wish to pay tribute to Dr W. Glaser, now with the Farrand Optical Company, New York, whose publications first made me aware of the power and elegance of the variational method in its application to electron optics.

P. A. S.

CAMBRIDGE
March 1955

PART I. STATIC ELECTRON OPTICS

CHAPTER 1

THE VARIATIONAL EQUATION

1.1. Introduction

The first part of this monograph deals with 'static' electron optics, i.e. the study of the optical properties of beams of electrons, or other charged particles, which are in *steady* motion so that, although the electrons making up a beam are in motion along their individual trajectories, the appearance of the beam as a whole does not vary in time.

In constructing a theory of static electron optics, one must take into account some or all of the following factors:

(*a*) the corpuscular properties of electrons,
(*b*) the wave properties of electrons,
(*c*) long-range interaction, and
(*d*) radiation reaction and short-range interaction.

Although the final examination of the performance of an electron-optical instrument will probably entail consideration of two or more of these factors, the initial investigation and basic design are frequently carried out on the assumption that (*a*) is predominant. The major part of the design of an electron microscope would entail this approximation, although it would be impossible to calculate the resolving power without consideration of (*b*). Similarly, the design of a cathode-ray tube can be based largely on the same approximation, but a more exact estimate of the performance for high-beam currents would necessitate the introduction of (*c*). It will be seen that (*d*) may always be neglected. Indeed, it is only if the first factor is dominant that one may expect to obtain image formation; diffraction effects, high-current densities and, of course, appreciable variation of the field during the time of transit of electrons, all tend to mar an image.

The theory of static electron optics which takes account of (*a*) and, if necessary, (*c*), but necessarily neglects (*b*) and (*d*), is known as *geometrical electron optics*, since, as will be seen in Chapter 2, much of the theory may be conveniently expressed in terms of geometrical concepts. Since the fields are static, no interest attaches to the motion of electrons with time but only to their spatial trajectories so that our study may be further restricted by the definition:

Geometrical electron optics is the study of the spatial trajectories and of the associated image formation of electrons moving in static electromagnetic fields, radiation reaction and short-range forces being neglected and dynamical laws taken in their classical form.

It is natural to regard the spatial trajectories as *electron rays*, and it will be shown in this chapter that, under the conditions laid down, electron rays satisfy a variational equation formally identical with Fermat's principle of light optics.‡

There are four observations concerning our definition which should be made at this point. The first is to explain that radiation reaction becomes important only in particle accelerators such as the betatron and electron-synchrotron where electrons of very high energies follow curved paths, and that short-range interactions, by which we mean forces arising from statistical fluctuation of the electromagnetic field due to 'granulation' of the beam,§ are never significant in electron-optical instruments, since the beams are not sufficiently dense.

The second point concerns the long-range interaction (*c*), by which we mean the Coulomb and Lorentz forces exerted by the beam, regarded now as a uniform fluid, upon its constituent elements. These are represented by the space-charge and space-current distributions, but if their effect is important the technical difficulty arises that the electron trajectories are determined by the electromagnetic field while the field is determined partly by the space charge which in turn depends on the trajectories. Hence although general rules, such as those to be established in Chapter 2, are implicitly valid even for high-density beams, explicit calculation of the effect of space charge will be considered only briefly in §4.5.

The third remark is that, although time itself is of no interest, it is essential to introduce time as independent variable if electrons

‡ The term 'light optics' is unfortunately both objectional and indispensable.

§ For an estimate of the forces due to 'granulation' see, for instance, *Microwave Electronics*, ed. G. B. Collins (McGraw-Hill, New York, 1948), pp. 221, 222. For a deeper analysis of the separation of Coulomb and Lorentz forces into 'short-range' and 'long-range' components, see D. Bohm and D. Pines, *Phys. Rev.* **82** (1951), 625–34; **85** (1952), 338–53.

are ever stationary along their paths, as they are in electron mirrors. It is not proposed to investigate electron mirrors‡ in this monograph, but, if it were, this investigation would fit more naturally into the second part, which deals with dynamic electron optics.

The fourth remark is that we have, by our definition, excluded from consideration certain calculations which are essential to the investigation of any electron-optical instrument, namely, the initial calculation of the electromagnetic field. However, the only general and practicable method for numerically calculating the fields of electron lenses appears to be the relaxation method,§ which it would be inappropriate to reproduce in these pages. Moreover, it is seldom that the fields are computed by numerical methods; they are more easily determined by means of an analogue computer such as the electrolytic tank|| or the more accurate resistance network.‡ In some cases it is even possible to measure the field directly.

1.2. Electron-optical units

Since we shall be concerned with only one type of charged particle,‡‡ it will be convenient to introduce units of field potentials so chosen that no physical constants appear explicitly in our formulation of the variational equation or, consequently, in the subsequent theoretical considerations.

However, let us first notice that since the trajectory of an electron depends, among other things, upon its energy on entering the field, the optical properties of a field should always be referred to a given monokinetic or—in optical terminology—*monochromatic* beam, i.e. a beam whose electrons all have the same energy on crossing an arbitrary equipotential surface of the electric field. Any departure from this condition in an electron-optical instrument will result in image defects which are classified as *chromatic aberration.* Since we are neglecting the short-range interaction of electrons, it is possible to consider electrons of various energies independently

‡ The theory of electron mirrors was first developed by A. Recknagel (*Z. Phys.* **104** (1937), 381–94), but a modern treatment is to be found in ref. (1).

§ R. V. Southwell, *Relaxation Methods in Theoretical Physics* (Oxford University Press, 1946).

|| See refs. (1) and (3).

‡ G. Liebmann, *Brit. J. Appl. Phys.* **1** (1950), 93–103.

‡‡ Except in § 5.5, where we deal with mass spectrographs.

so that chromatic aberration may be calculated within the framework of geometrical electron optics.

Let us now consider a monochromatic beam of electrons moving in an electric field whose scalar potential, measured in e.s.u., is ϕ^*. We may take account of the initial energy of the electrons, as well as of the energy which they acquire in the field, by adjusting ϕ^* so that *electrons are at rest at zero potential.* If $-e$ is the charge of an electron, measured in e.s.u., and if m_0 and m are the rest mass and relativistic mass, respectively,

$$e\phi^* = (m - m_0)\,c^2, \tag{1.2.1}$$

where c is the velocity of light; m_0, m, c and v, the speed of an electron, as measured in c.g.s. units.

If there were no electric field but only a magnetic field of strength H^*, measured in e.m.u., normal to the direction of motion of an electron of the beam, the radius of curvature R of the electron trajectory could be found by equating the centrifugal force mv^2/R to the Lorentz force evH^*/c. We see that $H^*R = p^*$, where p^*, defined by

$$p^* = mcv/e, \tag{1.2.2}$$

is a measure of the *scalar kinetic momentum.*

We shall see in the next section that the variational equation determining electron rays may be expressed in terms of the momentum p^* and the magnetic vector potential $\mathbf{A}^*$ only. The constants e, c and m_0 will therefore be eliminated from our calculations if they are eliminated from the relations between p^* and ϕ^* and between p^* and H^*. On using the well-known relation

$$m = m_0/\sqrt{(1 - (v/c)^2)}, \tag{1.2.3}$$

we find that the first of our relations becomes

$$p = \sqrt{(2\phi + \phi^2)}, \tag{1.2.4}$$

and the second retains the form $p = HR$ if we write

$$\phi = (e/m_0c^2)\,\phi^*, \quad p = (e/m_0c^2)\,p^*, \quad H = (e/m_0c^2)\,H^*. \tag{1.2.5}$$

Hence we should measure electric potential in units of e/m_0c^2 e.s.u. and magnetic field strength in units of e/m_0c^2 e.m.u.; thus:

Unit of electric potential = 511,200 volts,
Unit of magnetic field strength = 1,704 gauss.

Provided that related units are derived appropriately from these units, the usual relations between the electric and magnetic field vectors and the electric scalar and magnetic vector potentials remain valid:

$$\mathbf{E} = -\operatorname{grad}\phi, \quad \mathbf{H} = \operatorname{curl}\mathbf{A}. \tag{1.2.6}$$

Electron momentum, expressed as a magnetic quantity, will be measured in the same units as the magnetic potential (i.e. in units of 1704 gauss cm.).

The above units will be used in all subsequent calculations, but it should be noted that *one may at any time return to e.s.u. and e.m.u. by means of the formulae* (1.2.5).

Since our calculations have been based upon relativistic mechanics, the relation (1.2.4) is *relativistically correct*. However, if the beam energy is small compared with our unit of 511,200 volts, (1.2.4) may be approximated by

$$p = \sqrt{(2\phi)}, \tag{1.2.4a}$$

which is its *non-relativistic* form. Calculations will generally be relativistically correct.

Let us now consider how we should take into account the existence of a steady space charge of density ρ^*, measured in e.s.u., and a steady space current of density $\mathbf{j}^*$, measured in e.m.u. If we write

$$\rho = (4\pi e/m_0c^2)\rho^*, \quad \mathbf{j} = (4\pi e/m_0c^2)\mathbf{j}^*, \tag{1.2.7}$$

the inhomogeneous field equations take the simple forms

$$\nabla^2\phi = -\rho \tag{1.2.8}$$

and

$$\operatorname{curl}\operatorname{curl}\mathbf{A} = \mathbf{j}. \tag{1.2.9}$$

It is therefore proposed that we adopt the following units:

Unit of charge $= 4{\cdot}524 \times 10^{-8}$ *coulombs*,
Unit of current $= 1356$ *amperes*.

The unit of charge density may be expressed alternatively as

Unit of charge density $= 2{\cdot}834 \times 10^{11}$ *electronic charges/c.c.*

It may be established that if a beam of electrons has, at any point, energy ϕ, momentum p, space-charge density ρ and space-current density $\mathbf{j}$, then

$$\mathbf{j} = \left(\frac{p}{1+\phi}\right)\rho\mathbf{l}, \tag{1.2.10}$$

where $\mathbf{l}$ is the unit vector in the direction of motion.

The most important application of electron optics is, of course, to systems involving beams of electrons, but it may also be applied, with only minor modifications, to problems concerning beams of protons or ions. Since these particles carry a charge of the opposite sign to that of the electron, it is necessary to replace ϕ and $\mathbf{A}$ by $-\phi$ and $-\mathbf{A}$. The difference in specific charge—the ratio e/m_0—is reflected as the difference in the appropriate units; for instance, the units appropriate to proton beams are as follows: *unit of electric potential* $= 9{\cdot}391 \times 10^8$ *volts*; *unit of magnetic field strength* $= 3{\cdot}130 \times 10^6$ *gauss*; *unit of space charge* $= 8{\cdot}310 \times 10^{-5}$ *coulomb/c.c. or* $5{\cdot}187 \times 10^{14}$ *electronic charges/c.c.*; and *unit of current* $= 2{\cdot}491 \times 10^6$ *amperes*. An alternative is to make all measurements in e.s.u. and e.m.u. and to regard $\phi, p, \mathbf{H}, \rho$ and $\mathbf{j}$ as abbreviations for the expressions given in (1.2.5) and (1.2.7); this would be necessary in the study of mass spectrographs whose purpose is the determination of the specific charges of ions.

It is perhaps unnecessary to state that if one is considering the effect of the space charge of one beam of particles upon the optical properties of a second beam, the space charge and current of the first beam should be measured in units appropriate to the second, i.e. the 'focused', beam. However, it is worth noticing that the relation (1.2.10) would remain valid provided that one measures the energy and momentum of the first, i.e. the 'space charge', beam in units appropriate to that beam.

1.3. Derivation of the variational equation‡

We shall now proceed to derive from the principle of least action§ the variational equation which will be taken as the basis of geometrical electron optics.

If the velocity vector $\mathbf{v}$ has Cartesian components v_r, where $r = 1, 2, 3$, the principle of least action, as applied to a single electron, may be written as

$$\delta \int_A^B \mathbf{v} . \frac{\partial L}{\partial \mathbf{v}} \, dt = 0, \tag{1.3.1}$$

where L is the Lagrangian function, t is time and the notation $\partial/\partial \mathbf{v}$ is adopted for the vector operator with components $\partial/\partial v_r$. The variational operator refers, in this case, to all variations of the trajectory which leave the terminal points A and B and the function $\mathbf{v}.(\partial L/\partial \mathbf{v}) - L$ invariant. It should be noted that (1.3.1) *and all*

‡ Since it would seem unreasonable to grace every variational formulation of the ray equations and of the equations of motion with the title of 'principle', such a formulation will normally be referred to as a *variational equation*.

§ See ref. (12), p. 207, eq. (65.9).

subsequent equations involving the symbol δ are exact only to the first order in the increments due to δ.

If, for the purposes of this section, the unit of time is so chosen that the velocity of light is unity, the Lagrangian function for an electron‡ is

$$L = 1 - \sqrt{(1-v^2)} + \phi - \mathbf{v}.\mathbf{A}. \tag{1.3.2}$$

The invariance under the variational operation of the function $\mathbf{v}.(\partial L/\partial \mathbf{v}) - L$ now leads to the invariance of the equation

$$1 + \phi = 1/\sqrt{(1-v^2)}. \tag{1.3.3}$$

If s measures arc length along the trajectory, $ds = v\,dt$, so that the time integral of (1.3.1) may be replaced by a line integral. If we now eliminate v by means of (1.3.3), we obtain as the variational equation defining the rays of a given monochromatic electron beam in a static electromagnetic field

$$\delta \int_A^B \{\sqrt{(2\phi + \phi^2)} - \mathbf{l}.\mathbf{A}\}\, ds = 0, \tag{1.3.4}$$

where $\mathbf{l}$ is a unit vector along the tangent to the trajectory and δ now refers to all variations of the path which leave the terminal points invariant.

Equation (1.3.4) will be adopted as the basis of our theory of geometrical electron optics.

It is interesting (though quite irrelevant) to consider the range of values of ϕ for which (1.3.4) is physically significant. It is clearly necessary, for practical electron optics, that $\phi \geqslant 0$. For small negative values of ϕ, the radical becomes imaginary so that, in classical mechanics, $\phi = 0$ represents a lower impassable boundary. However, let us suppose that by some process outside the scope of classical mechanics ϕ were depressed below -2. Let us write

$$\phi = \phi^* - 2, \tag{1.3.5}$$

and assume that ϕ^* is negative. When ϕ is positive the radical takes a positive value; when ϕ is negative let us give the radical a negative value. We then find that (1.3.4) may be written as

$$\delta \int_A^B \{\sqrt{(-2\phi^* + \phi^{*2})} + \mathbf{l}.\mathbf{A}\}\, ds = 0. \tag{1.3.6}$$

If we now refer to § 2 and find the variational equation which applies to 'ions' whose specific charge is the same as that of

‡ See ref. (12), p. 349, eq. (99.6).

electrons, we obtain exactly the form (1.3.6). This shows that *when the energy of the electron is reduced by more than twice its rest energy, it behaves as a particle with the same mass but positive charge.* This is in agreement with Dirac's positron theory.‡

Let us now apply the equation (1.3.4) to the simple problem of slow electrons moving in a strong magnetic field. In this case we can neglect the first term of (1.3.4) so that it becomes

$$\delta \int_A^B \mathbf{A} . d\mathbf{s} = 0. \tag{1.3.7}$$

Now the integral (1.3.7), taken around a closed circuit, gives the magnetic flux enclosed by the circuit, as is readily seen from (1.2.6) and application of Stokes's theorem. The equation (1.3.7) therefore states that the closed circuit formed by a ray and an arbitrary slight displacement of the ray embraces no magnetic flux. This is possible only if there is no component of field strength normal to the ray, so that slow electrons in a strong magnetic field must follow the lines of field strength.

If we take into account small but finite energy of the electrons, their paths will differ slightly from the lines of field strength. In any small neighbourhood their motion must resemble the motion of electrons in a uniform magnetic field. The latter, as we shall see later, § is a helix so that *slow electrons in a strong magnetic field move in helices about the lines of field strength.*

The variational equation is expressed in (1.3.4) explicitly in terms of the potentials, but it will be convenient for many general discussions to write it in the shorter form

$$\delta \int_A^B \{p - \mathbf{1} . \mathbf{A}\} ds = 0, \tag{1.3.8}$$

where p, the scalar momentum of the beam, is given by (1.2.4). We may observe, incidentally, that for a purely magnetic field, for which p is a constant, it is not necessary to measure p and $\mathbf{A}$ in electron-optical units although they must be measured in the same units; if the field is purely electric, the units may be left arbitrary only if the treatment is non-relativistic. We may also note that the *direction* of the ray enters only by way of the unit vector $\mathbf{1}$ and that

‡ P. A. M. Dirac, *Quantum Mechanics* (Oxford University Press, 3rd ed. 1947), p. 272.

§ See pp. 19, 20.

reversal of $\mathbf{l}$ is equivalent to a change in sign of $\mathbf{A}$. It follows that *electron rays are reversible only if the field is purely electric*; if the field is partly or wholly magnetic, rays can be reversed only if the sense of the magnetic field is reversed.

Let us now write (1.3.8) in the form

$$\delta\int_A^B n\,ds = 0, \tag{1.3.9}$$

where the function $n(\mathbf{x}, \mathbf{l})$ is defined by

$$n = p - \mathbf{l}.\mathbf{A}. \tag{1.3.10}$$

It is now obvious that electron rays possess optical properties, for (1.3.9) is formally identical with Fermat's principle of light optics. We shall, by analogy, call the quantity n defined by (1.3.10) the *refractive index* of the field. It is to be remembered that, just as the refractive index of glass depends on the colour of the light beam, so the electron-optical refractive index depends implicitly upon the energy with which the electron beam enters the field.

It is well known that the electric scalar potential and the magnetic vector potential are to some extent arbitrary. We have made use of the indeterminacy of the former to combine the initial energy of the beam with the potential of the field, but the magnetic potential $\mathbf{A}$ is still arbitrary in that we may add to it the gradient of an arbitrary scalar distribution $\chi(\mathbf{x})$, say. Let us investigate briefly the consequences of this indeterminacy in $\mathbf{A}$.

If we change $\mathbf{A}$ to $\mathbf{A}+\operatorname{grad}\chi$, (1.3.8) becomes

$$\delta\int_A^B \{p - \mathbf{l}.\mathbf{A}\}\,ds + \delta\chi_a - \delta\chi_b = 0, \tag{1.3.11}$$

where the suffix a or b will generally denote that a function is evaluated at A or B, respectively. Since δ refers to variations for which $\delta\mathbf{x}_a = \delta\mathbf{x}_b = 0$, the integrated terms of (1.3.11) vanish. Hence *the variational equations formed from equivalent potential distributions are themselves equivalent.*

We see from (1.3.10) that, at any point, the refractive index is *isotropic* or *anisotropic*—i.e. it does not depend upon or depends upon the direction vector—according as the magnetic potential vanishes or does not vanish, respectively, at that point. It is obvious that, by replacing a distribution $\mathbf{A}$ by a distribution $\mathbf{A}+\operatorname{grad}\chi$, where χ is suitably chosen, *the refractive index may be made isotropic at any finite number of points* so that the notion of isotropy *at a point* is without physical significance.

Let us now consider the refractive index over a surface (or finite number of surfaces). If we resolve the vector potential at each point of this surface into components $\mathbf{A}_n$ and $\mathbf{A}_t$ which are normal and tangential, respectively, to the surface at that point, there will clearly be no difficulty in eliminating $\mathbf{A}_n$ by adding to $\mathbf{A}$ the gradient of some potential distribution. However, it will not be possible simultaneously to eliminate $\mathbf{A}_t$ unless it is possible so to arrange the distribution of χ in the surface that $\mathbf{A}_t + \text{grad}_t\chi$ vanishes, where $\text{grad}_t\chi$ is the tangential component of $\text{grad}\,\chi$. By considering integrals along arbitrary closed curves lying in the surface and applying Stokes's theorem, we find that the necessary and sufficient condition for this to be possible is that the normal component of $\text{curl}\,\mathbf{A}$ vanishes. Hence *the refractive index may be made isotropic at all points of a given surface if and only if the normal component of the magnetic field strength is zero at all points of the surface.*

The necessary and sufficient condition that, given a vector potential distribution $\mathbf{A}$, we may find a scalar distribution χ such that $\mathbf{A} + \text{grad}\,\chi = 0$ throughout a given volume is that $\text{curl}\,\mathbf{A}$ vanishes throughout the volume. It follows that *the refractive index may be made isotropic over a given volume if and only if the magnetic field strength vanishes throughout the volume.* If we say that a *field* is isotropic at a point if it is possible to make the refractive index isotropic throughout a small volume containing the point and anisotropic otherwise, we see that *an isotropic field is one which is purely electric whereas a field which is partly or wholly magnetic is anisotropic.*

1.4. The electrostatic lens

Now that the variational equation has been established, it will be instructive to make a preliminary investigation of one or two typical electron-optical systems in order to determine how the equation may be applied to certain simple but important calculations, and in order to decide what further calculations we shall ultimately wish to make. In this section we shall establish the paraxial imaging properties of an electric field of rotational symmetry such as is produced by an 'electrostatic electron lens'. An electrostatic lens may comprise any rotationally symmetrical electrode assembly but is usually designed to produce a strong field of small dimensions. The construction of a typical electrostatic lens is shown in figure 1.1.‡

Let us adopt both rectangular co-ordinates (x_1, x_2, z) and cylindrical co-ordinates (z, r, θ) so that

$$x_1 + \mathrm{i}x_2 = r\,\mathrm{e}^{\mathrm{i}\theta}, \tag{1.4.1}$$

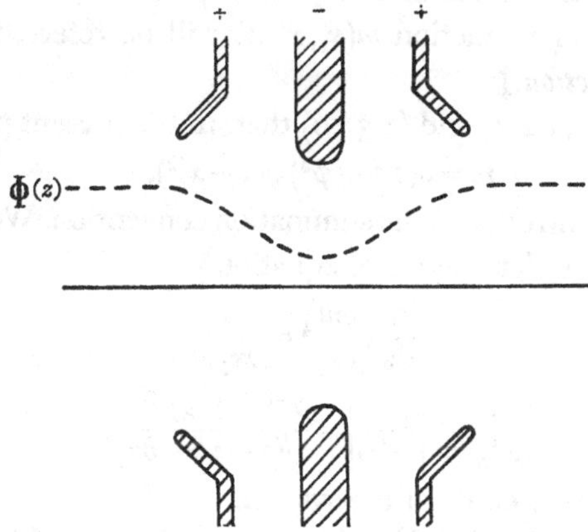

Fig. 1.1. Typical electrostatic lens ('three-electrode' lens).

and assume that the system is rotationally symmetrical about the z-axis. If we also specify that there is no space charge in the neighbourhood of the axis, the field equation (1.2.8) becomes§

$$\left\{\frac{\partial^2}{\partial z^2} + \frac{\partial^2}{\partial r^2} + \frac{1}{r}\frac{\partial}{\partial r}\right\}\phi = 0. \tag{1.4.2}$$

If ϕ is expressed as a power series in r^2, the relations between the coefficients of this series may be determined from (1.4.2). In this way we obtain the expansion

$$\phi(z, r) = \Phi - \tfrac{1}{4}r^2\Phi'' + \tfrac{1}{64}r^4\Phi^{\mathrm{iv}} - \ldots, \tag{1.4.3}$$

where $\Phi(z)$ is the value of ϕ upon the z-axis and primes will be used to denote total differentiation with respect to z.

‡ For enumeration of the various types of electrostatic lenses, see refs. (1), (3) and (15).

§ This equation may be derived by means of the standard formulae for vector operators in curvilinear co-ordinates: see, for instance, J. C. Slater and N. H. Frank, *Electromagnetism* (McGraw-Hill, New York, 1947), Appendix 4. The derivation is carried out in detail in refs. (1), (2), (3) and (4).

For most calculations of the properties of electron-optical instruments, it is convenient to rewrite (1.3.9) as

$$\delta \int_A^B m\,\mathrm{d}z = 0, \tag{1.4.4}$$

where

$$m = n\frac{\mathrm{d}s}{\mathrm{d}z} \tag{1.4.5}$$

for if we treat z as independent variable and x_i, where i takes the values 1 and 2, as dependent variables, we shall be in a position to investigate the intersections of rays with planes normal to the axis of symmetry. The function $m(x_i, x_j', z)$ will be referred to as the *variational function.*‡

We see from (1.2.4) and (1.3.10) that, in the present problem,

$$m = \sqrt{(2\phi + \phi^2)}\sqrt{(1 + x_i'^2)}, \tag{1.4.6}$$

where we have introduced the summation convention. We find from (1.4.6) that the Euler-Lagrange equation§

$$\frac{\mathrm{d}}{\mathrm{d}z}\left(\frac{\partial m}{\partial x_i'}\right) = \frac{\partial m}{\partial x_i} \tag{1.4.7}$$

becomes

$$\frac{\mathrm{d}}{\mathrm{d}z}\left\{\frac{p x_i'}{\sqrt{(1 + x_j'^2)}}\right\} = \frac{p^{-1}(1+\phi)}{\sqrt{(1+x_j'^2)}}\frac{\partial \phi}{\partial x_i}, \tag{1.4.8}$$

which is the *ray equation* of the system.

Since the equation (1.4.8) holds for *any* electric field, we cannot expect it to bring to light the image-forming properties peculiar to fields of rotational symmetry. Let us therefore substitute the expansion (1.4.3) into (1.4.6); we may then expand the variational function in the form

$$m = m^{(0)} + m^{(2)} + m^{(4)} + \dots, \tag{1.4.9}$$

where $m^{(s)}$ denotes a homogeneous polynomial of the sth degree in x_i and x_i'. Since $m^{(0)}$ does not involve the dependent variables, it cannot contribute to the ray equation (1.4.7) and so may be omitted. The variational function then tends to $m^{(2)}$ as x_i and x_i' become small.

If we assume that the system does not behave as a mirror,‖ x_i' becomes small as x_i becomes small, so that, for narrow beams

‡ The integrand of any variational equation may be referred to as a 'Lagrangian function' or 'Lagrangian', but, in dynamics, 'Lagrangian' usually denotes the integrand in a particular formulation of Hamilton's principle. In order to avoid confusion, the integrand of any variational equation will be referred to as a *variational function.*

§ See p. 64, eqs. (3.2.5) and (3.2.6).

‖ In electron mirrors, x_i' is infinite at the point of 'reflexion', so that z may not be used as independent variable. See ref. (1).

moving close to the axis, we may with little error replace m by $m^{(2)}$. The approximation entailed in neglecting all terms of the variational function of higher order than the second is therefore known as the *paraxial approximation*. We find that

$$m^{(2)} = \tfrac{1}{2}\mathsf{p}x_i'^2 - \tfrac{1}{4}\mathsf{p}^{-1}(1+\Phi)\Phi'' x_i^2, \qquad (1.4.10)$$

where
$$\mathsf{p} = \sqrt{(2\Phi + \Phi^2)}, \qquad (1.4.11)$$

i.e. $\mathsf{p}(z)$ is the value of p upon the axis. The Euler-Lagrange equation derivable from (1.4.10),

$$\frac{\mathrm{d}}{\mathrm{d}z}\left\{\mathsf{p}\frac{\mathrm{d}x_i}{\mathrm{d}z}\right\} + \tfrac{1}{2}\mathsf{p}^{-1}(1+\Phi)\Phi'' x_i = 0, \qquad (1.4.12)$$

is the *paraxial ray equation* for electric fields of rotational symmetry. It is apparent that the paraxial ray equation may also be obtained by picking out the part of the complete ray equation (in this case (1.4.8)) which is of the first order in the dependent or 'off-axis' co-ordinates.‡

Since (1.4.12) is a second-order linear differential equation, its general solution may be expressed as a combination of any two linearly independent particular solutions. Let us consider the imaging of an object plane $z = z_o$ upon an image plane $z = z_b$, and let us introduce an aperture plane $z = z_a$ at which the cross-section of the beam may be defined by an aperture of 'stop'. Then if $g(z)$ and $h(z)$ are solutions of (1.4.12) satisfying the boundary conditions (fig. 1.2)

$$g_o = 1, \quad g_a = 0, \qquad h_o = 0, \quad h_a = 1, \qquad (1.4.13)$$

where $g_o = g(z_o)$, etc., the general solution of (1.4.12) may be written as

$$x_i(z) = x_{io}g(z) + x_{ia}h(z). \qquad (1.4.14)$$

In a two-dimensional picture, the rays defined by $x = g(z)$ and $x = h(z)$ are sometimes referred to as the *principal rays* of the system (fig. 1.2).

It is clear from (1.4.14) that the condition that the planes with co-ordinates z_o and z_b should be paraxially imaged is that

$$h_b = 0 \qquad (1.4.15)$$

and that the magnification is then given by

$$M = g_b. \qquad (1.4.16)$$

‡ The advantage of the former approach will become apparent later when we also pick out from the variational function the terms responsible for various aberrations.

It follows from (1.4.15) that in specifying (1.4.13) we are implicitly specifying that the object and aperture planes are *non-conjugate*, i.e. that the object plane is not imaged upon the aperture plane. Moreover, if (1.4.15) and (1.4.16) hold for one choice of z_a, they hold for any other choice.

If we define k by
$$k = \mathsf{p}(g'h - gh'), \tag{1.4.17}$$
it is easy to verify from (1.4.12) that $k' = 0$ so that k is independent of z. It will be found that k is a form of the *Lagrange invariant*, which will be introduced in the next chapter.‡ It follows from (1.4.13) and (1.4.15) that
$$k = -\mathsf{p}_o h'_o, \quad k = \mathsf{p}_a g'_a \quad \text{and} \quad k = -\mathsf{p}_b g_b h'_b. \tag{1.4.18}$$
To the paraxial approximation, the angle at which a ray cuts the axis is given by
$$\omega = \mathrm{d}r/\mathrm{d}z. \tag{1.4.19}$$

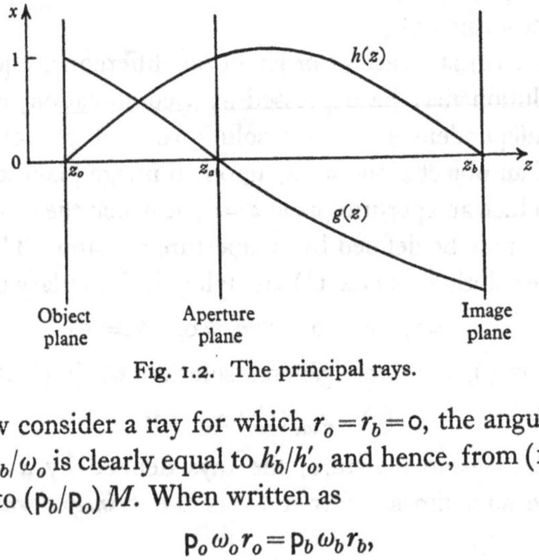

Fig. 1.2. The principal rays.

If we now consider a ray for which $r_o = r_b = 0$, the angular magnification ω_b/ω_o is clearly equal to h'_b/h'_o, and hence, from (1.4.16) and (1.4.18), to $(\mathsf{p}_b/\mathsf{p}_o)M$. When written as
$$\mathsf{p}_o \omega_o r_o = \mathsf{p}_b \omega_b r_b, \tag{1.4.20}$$
this rule is recognized as the *Lagrange relation*.

The concepts of cardinal points and focal lengths will be investigated in § 3·4, but it is proposed that we here assume that a pair of focal lengths f_o and f_b may be ascribed to an electrostatic lens whose field vanishes over all but a finite length of the axis and attempt to calculate their values assuming the field to be weak.

‡ See eqs. (2.3.3) and (3.3.37). This invariant is known by many names, but the one we have adopted seems most appropriate to optics.

If the suffixes *o* and *b* now denote the values of quantities anywhere in the object and image 'spaces' (i.e. the regions outside the lens associated with the object and image), respectively, and if we define $s(z)$ and $t(z)$ (fig. 1.3) to be two solutions of (1.4.12) satisfying the boundary conditions

$$s_o = 1, \quad s'_o = 0, \qquad t_b = 1, \quad t'_b = 0, \tag{1.4.21}$$

then

$$f_o = 1/t'_o, \quad f_b = -1/s'_b. \tag{1.4.22}$$

The invariance of the combination $p(s't - st')$ shows at once that

$$f_b/f_o = p_b/p_o. \tag{1.4.23}$$

Now we see from (1.4.12) and (1.4.21) that

$$p_b s'_b = -\frac{1}{2}\int_{-\infty}^{\infty} p^{-1}(1+\Phi)\,\Phi'' s\,dz, \tag{1.4.24}$$

where it is permissible to extend the range of integration over the whole z-axis. It is reasonable to seek to obtain from (1.4.24) a formula for the focal length of a weak lens by putting $s(z) = 1$ in

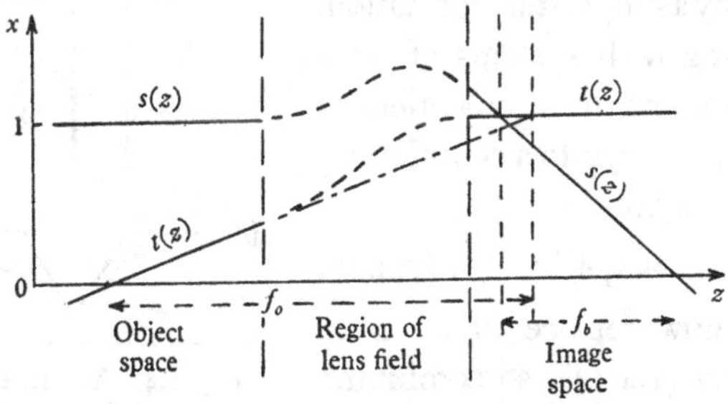

Fig. 1.3. Definition of the focal lengths.

the integrand; this should be a good approximation if rays are deflected only slightly by the field. However, an integration by parts makes it possible to replace the integrand of (1.4.24), which is of the first order in the strength of the lens field, by $\frac{1}{2}p^{-3}\Phi'^2$, which is of the *second* order in the field strength. Consequently it is not possible to obtain in this way a formula for the focal length valid for weak fields without estimating the second-order contribution to the integral (1.4.24) arising from the part of $s(z)$ which is linear in the lens field strength.‡ A simpler derivation of the required formula will be given later.

‡ Several authors have overlooked this point and so obtained an erroneous formula for the thin-lens approximation to the focal length of electrostatic electron lenses.

In integrating (1.4.24) (with $s(z)=1$) by parts, we ignored the integrated term $-\frac{1}{2}\mathrm{p}^{-1}(1+\Phi)\Phi'$, since the field was assumed to vanish outside the region of the lens. However, if we consider an 'aperture lens' which is comprised of a plane electrode with a small circular aperture, on the object side of which the field strength is Φ_1' and on the image side Φ_2', the integrated term gives the most important contribution. Neglecting the change in Φ, we obtain

$$1/f=\tfrac{1}{2}\mathrm{p}^{-2}(1+\Phi)(\Phi_2'-\Phi_1'). \tag{1.4.25}$$

This formula remains valid if the field is limited by foils or fine-mesh grids situated near the aperture electrode (fig. 1.4).

The ray equation (1.4.12) is not particularly convenient for numerical calculation since, when the first term is expanded, it involves a first-order differentiation. This defect may be overcome by a simple transformation.

In dealing with systems of rotational symmetry, it is convenient to combine x_1 and x_2 into a complex coordinate by writing

$$u=x_1+\mathrm{i}x_2. \tag{1.4.26}$$

We may now replace x_i by u in (1.4.12) and (1.4.14), so combining each pair of equations. If we now write

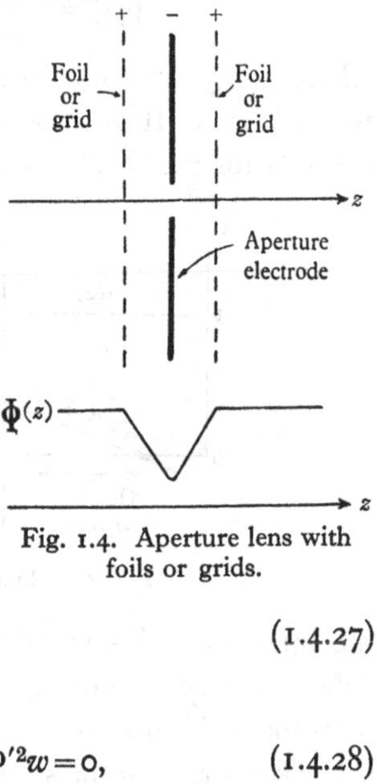

Fig. 1.4. Aperture lens with foils or grids.

$$u=\mathrm{p}^{-\frac{1}{2}}w, \tag{1.4.27}$$

the equation (1.4.12) reduces to

$$\frac{\mathrm{d}^2w}{\mathrm{d}z^2}+\tfrac{1}{4}\mathrm{p}^{-4}(3+\mathrm{p}^2)\Phi'^2w=0, \tag{1.4.28}$$

of which the non-relativistic form is

$$\frac{\mathrm{d}^2w}{\mathrm{d}z^2}+\frac{3}{16}\left(\frac{\Phi'}{\Phi}\right)^2w=0; \tag{1.4.28a}$$

these equations are much more convenient for computation.‡ A further advantage of (1.4.28) over (1.4.12) is that it involves Φ' instead of Φ''; the former can be obtained more accurately than the

‡ See D. R. Hartree, *Numerical Analysis* (Oxford University Press, 1952), pp. 126 and 136, and J. C. Burfoot, *Brit. J. Appl. Phys.* 3 (1952), 22–4.

latter from an experimental determination of $\Phi(z)$. It is interesting to note from (1.4.28) that any electric field of finite extension is *convergent* in the sense that any ray which is parallel to the axis in the object space and does not cross the axis will be directed towards the axis in the image space.

Since the coefficient of w in (1.4.28) is of the second order in the field strength, the difficulty which we met in trying to calculate the focal length from (1.4.12) will not be repeated. If $s(z)$ and $t(z)$ are now to be solutions of (1.4.28), it follows from (1.4.27) that (1.4.21) and (1.4.22) should be modified to read

$$s_o = \mathrm{P}_o^{\frac{1}{4}}, \quad s_o' = 0, \qquad t_b = \mathrm{P}_b^{\frac{1}{4}}, \quad t_b' = 0 \tag{1.4.29}$$

and

$$f_o = \mathrm{P}_o^{\frac{1}{4}}(1/t_o'), \quad f_b = -\mathrm{P}_b^{\frac{1}{4}}(1/s_b'). \tag{1.4.30}$$

The invariance of $s't - st'$ now leads to (1.4.23). If we now calculate s_o' from the integrated form of (1.4.28), replacing $s(z)$ in the integrand by unity, we obtain

$$\frac{\mathrm{P}_o}{f_o} = \frac{\mathrm{P}_b}{f_b} = \tfrac{1}{4}(\mathrm{P}_o\mathrm{P}_b)^{\frac{1}{4}} \int_{-\infty}^{\infty} \mathrm{P}^{-\frac{7}{4}}(3+\mathrm{P}^2)\,\Phi'^2\,dz. \tag{1.4.31}$$

The approximation which we have adduced in (1.4.31) is known sometimes as the 'weak-lens approximation', but more often as the 'thin-lens approximation', since the criterion for the applicability of this approximation is that the lens should be short in comparison with the focal length.

Let us consider, as a simple example, the approximate calculation of the focal length of a 'bipotential' lens such as is formed by two coaxial cylindrical electrodes maintained at different potentials. An idealized form of the axial potential distribution, which is accurate if the diameters of the cylinders are reduced to zero and the cylinders are given infinite plane faces, is given by

$$\left.\begin{aligned} &\Phi(z) = \Phi_1 && (z < -\tfrac{1}{2}d), \\ &\Phi(z) = \tfrac{1}{2}(\Phi_1+\Phi_2) + (\Phi_2-\Phi_1)\frac{z}{d} && (-\tfrac{1}{2}d < z < \tfrac{1}{2}d), \\ &\Phi(z) = \Phi_2 && (z > \tfrac{1}{2}d). \end{aligned}\right\} \tag{1.4.32}$$

If the non-relativistic form of the formula (1.4.31) is evaluated for this field, we obtain the formula

$$\frac{\mathrm{P}_1}{f_o} = \frac{\mathrm{P}_2}{f_b} = \tfrac{3}{16}(\mathrm{P}_1\mathrm{P}_2)^{\frac{1}{4}} \frac{(\Phi_2-\Phi_1)^2}{\Phi_1\Phi_2}\frac{1}{d}, \tag{1.4.33}$$

which may be compared with the exact formulae which will be established as (3.4.32). This comparison shows that (1.4.33) gives values of $1/f$ which are too large by a factor of about $\frac{1}{2}(d/f)$ so that even if $f = 5d$, the formula is still in error by a factor of 10 %. It follows that the thin-lens approximation, when applied to electrostatic lenses, can be relied on only to give 'orders of magnitude'.

1.5. The magnetic lens

It is proposed that we now investigate briefly the electron-optical properties of magnetic fields of rotational symmetry. Although, as we shall see, any such field possesses lens properties, magnetic lenses of the type used in electron microscopes are designed to produce intense fields of small dimensions.‡ The construction of a typical lens is indicated in fig. 1.5.

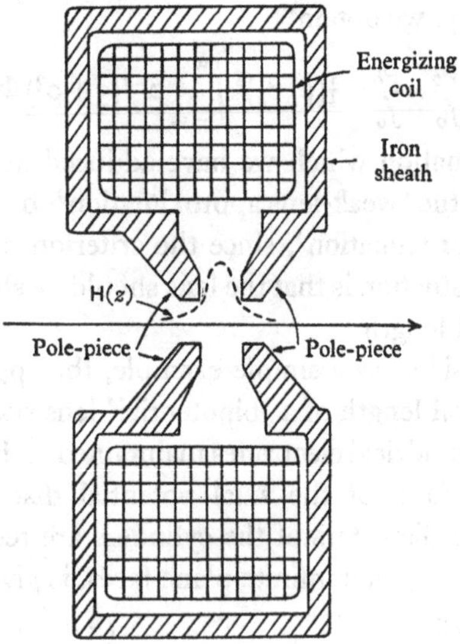

Fig. 1.5. Typical magnetic lens (iron-clad objective lens).

In terms of cylindrical co-ordinates, a magnetic field of rotational symmetry about the z-axis may be represented by a vector potential with only one non-zero component, $A_\theta(z, r)$. Assuming

‡ The reader is referred to refs. (1) and (3) for details of the construction of magnetic electron lenses.

that there are no space currents, we find that (1.2.9) gives the field equation‡

$$\left\{\frac{\partial^2}{\partial z^2}+\frac{\partial^2}{\partial r^2}+\frac{1}{r}\frac{\partial}{\partial r}-\frac{1}{r^2}\right\}A_\theta=0. \qquad (1.5.1)$$

It follows from (1.5.1) that A_θ may be expanded as

$$A_\theta(z,r)=\tfrac{1}{2}r\mathsf{H}-\tfrac{1}{16}r^3\mathsf{H}''+\ldots, \qquad (1.5.2)$$

where $\mathsf{H}(z)$ denotes $H_z(z,0)$, the value of the magnetic field strength on the z-axis, for (1.5.2) satisfies (1.5.1) and the line integral $2\pi rA_\theta$ tends to the surface integral πr^2H_z as r tends to zero.

The variational equation may again be written in the form (1.4.4), but we now find from (1.3.10) and (1.4.5) that

$$m=p\sqrt{(1+r'^2+r^2\theta'^2)}-r\theta'A_\theta. \qquad (1.5.3)$$

We should now find the Euler-Lagrange equations of the function (1.5.3). Since m does not depend explicitly on θ, the θ equation (which is obtained by replacing x_i in (1.4.7) by θ) may be integrated to give

$$pr^2\theta'(1+r'^2+r^2\theta'^2)^{-\frac{1}{2}}-rA_\theta=C, \qquad (1.5.4)$$

where C will be a constant for any ray. It is clear that C determines the 'skewness' of a ray for $C=0$ if a ray intersects the axis anywhere along its length. Equation (1.5.4) may be rearranged to give θ' in terms of z, r, r' and C:

$$\{p^2-(A_\theta+C/r)^2\}^{\frac{1}{2}}r\theta'=(1+r'^2)^{\frac{1}{2}}(A_\theta+C/r). \qquad (1.5.5)$$

The r equation derivable from (1.5.3) is

$$\frac{d}{dz}\{pr'(1+r'^2+r^2\theta'^2)^{-\frac{1}{2}}\}=pr\theta'^2(1+r'^2+r^2\theta'^2)^{-\frac{1}{2}}-\theta'(A_\theta+r\,\partial A_\theta/\partial r) \qquad (1.5.6)$$

which, when combined with (1.5.5), reduces to

$$\frac{d}{dz}\left\{\frac{r'\sqrt{(p^2-(A_\theta+C/r)^2)}}{\sqrt{(1+r'^2)}}\right\} = -\frac{(A_\theta+C/r)(\partial A_\theta/\partial r-C/r^2)}{\sqrt{(p^2-(A_\theta+C/r)^2)}}\sqrt{(1+r'^2)}. \qquad (1.5.7)$$

Equation (1.5.7) determines the motion in the z-r or 'meridional' plane of an assembly of rays with a given value of C. The same equation is obtained by postulating that the meridional projections of rays are determined by a 'meridional refractive index'

$$\sqrt{(p^2-(A_\theta+C/r)^2)};$$

‡ See p. 11, n. §.

it will be shown in § 2.7 that this same formula is valid even if the field is partly or wholly electric.

We may verify from (1.5.5) and (1.5.7) that in a uniform magnetic field all rays are helices. If the field is parallel to the z-axis and of strength H, $A_\theta = \frac{1}{2}r\mathsf{H}$ so that (1.5.5) and (1.5.7) admit solutions with $r = \text{const.}$ and $\theta' = \text{const.}$ provided that

$$\theta'^2(p^2/\mathsf{H}^2 - r^2) = 1. \tag{1.5.8}$$

Such a helix may be made to fit any initial conditions (corresponding to the position and direction of motion of an electron) by translation of the co-ordinate frame and adjustment of the pitch of the helix.

Let us now discuss the paraxial properties of magnetic fields of rotational symmetry. If (1.5.2) is substituted in (1.5.3) we again find that m may be expanded in the form (1.4.9) and that, in particular,

$$m^{(2)} = \tfrac{1}{2}p(r'^2 + r^2\theta'^2) - \tfrac{1}{2}r^2\theta'\mathsf{H}. \tag{1.5.9}$$

However, if (1.5.9) is expressed in terms of x_i and x_i' the Euler-Lagrange equations do not 'separate' in the sense that each equation involves only x_1 or x_2. This fact may be explained by reference to (1.5.5) which shows that, to the paraxial approximation, non-skew rays (for which $C = 0$) rotate about the z-axis by $\frac{1}{2}p^{-1}\mathsf{H}$ radians per unit length.

This suggests that, in terms of complex co-ordinates, we effect the transformation

$$u = v\,e^{i\chi}, \tag{1.5.10}$$

where $\chi(z)$ satisfies

$$\frac{d\chi}{dz} = \tfrac{1}{2}p^{-1}\mathsf{H}. \tag{1.5.11}$$

We then find, from (1.4.1), (1.4.26), (1.5.9), (1.5.10) and (1.5.11), that

$$m^{(2)} = \tfrac{1}{2}p\bar{v}'v' - \tfrac{1}{8}p^{-2}\mathsf{H}^2\bar{v}v, \tag{1.5.12}$$

from which we obtain the paraxial ray equation

$$\frac{d^2v}{dz^2} + \tfrac{1}{4}p^{-2}\mathsf{H}^2 v = 0, \tag{1.5.13}$$

which is in 'separated' or, to adopt the appropriate optical term, 'orthogonal' form, since it does not involve $\bar{v}$.

We see that, since (1.5.13) is of the same form as (1.4.12), magnetic fields of rotational symmetry possess paraxial imaging properties although the focusing effect is coupled with a rotation of rays about the axis. Equation (1.5.13) is already in the form of (1.4.28) which demonstrates that *magnetic lenses are always convergent.* Starting from (1.5.13) instead of (1.4.12), we may retrace the steps (1.4.13) to (1.4.20); the only difference is that p is now a constant. If we also proceed to calculate the focal lengths of a weak magnetic lens to the approximation adopted in the derivation of (1.4.25) and (1.4.31), we obtain the formula

$$\frac{1}{f_o}=\frac{1}{f_b}=\tfrac{1}{4}p^{-2}\int_{-\infty}^{\infty}\mathsf{H}^2\,dz, \tag{1.5.14}$$

which was first given by Busch. It is to be expected from (1.4.23) that, quite generally, the two focal lengths of a purely magnetic lens are equal.

It is worth considering an example of the application of (1.5.14) in order to appraise its accuracy. Let us suppose that $\mathsf{H}(z)$ has a constant value over a length l and vanishes outside this region; such a field would be produced approximately by a long narrow solenoid. Clearly (1.5.14) leads to

$$1/f=\tfrac{1}{4}p^{-2}\mathsf{H}^2 l. \tag{1.5.15}$$

Let us now calculate the focal length exactly by finding $s(z)$ which satisfies (1.5.13) and (1.4.21). If the field lies between $z=0$ and $z=l$,

$$s(z)=\cos\left(\tfrac{1}{2}p^{-1}\mathsf{H}z\right)\quad(0<z<l), \tag{1.5.16}$$

so that, from (1.4.22), we obtain the strict formula

$$1/f=\tfrac{1}{2}p^{-1}\mathsf{H}\sin\left(\tfrac{1}{2}p^{-1}\mathsf{H}l\right). \tag{1.5.17}$$

It is easy to see from (1.5.17) that, for weak lenses, (1.5.15) is in error by a factor of about $\tfrac{1}{6}(l/f)$, so that if $f=5l$ the error is about 3%. It follows that the thin-lens approximation gives better results for magnetic lenses than for comparable electrostatic lenses.

Another interesting example is the field form

$$\mathsf{H}(z)=\mathsf{H}_0\left(1+(z/a)^2\right)^{-\frac{3}{2}}, \tag{1.5.18}$$

which is exactly the field of a single turn of wire and approximately the field met in electron-microscope lenses. Evaluation of (1.5.14) leads to the formula

$$\frac{1}{f}=\frac{3\pi}{32}p^{-2}\mathsf{H}_0^2 a. \tag{1.5.19}$$

Suppose that the field of a lens has the shape of (1.5.18) with $a = 0{\cdot}25$ cm. and a maximum field strength of 2000 gauss, and that the lens focuses a beam of 50,000 volts energy. We see from § 1.2 that, in electron-optical units, $\mathsf{H}_0 = 1{\cdot}173$ and $\phi = 0{\cdot}0978$, so that, from (1.2.4), $p = 0{\cdot}453$. The formula (1.5.19) now gives an estimate of 1·27 cm. for the focal length, but we must expect this to be in error (short) by 5 % or more.

Now that we have seen how the variational equation may be made to yield either the exact ray equation or the paraxial ray equation of a system, let us decide what further calculations our theory should enable us to perform.

It has been observed that whereas the exact ray equation of a system of rotational symmetry does not guarantee image formation, the paraxial ray equation does. This means that an electric or magnetic field of rotational symmetry will produce a clear image of an object provided both the object and the aperture are sufficiently small. It is therefore necessary that we should be able to estimate the degree by which, and the manner in which, the image of a system with object and aperture of finite dimensions differs from its paraxial ideal. This may be achieved by calculating the *geometrical aberrations*‡ which are the modifications in image structure due to the terms $m^{(4)}$, $m^{(6)}$, etc., of the expansion (1.4.9) which were neglected in the paraxial calculations. The most important of these aberrations derive from $m^{(4)}$ and are known as the *third-order aberrations*, since their magnitude varies cubically with the magnitude of the off-axis co-ordinates.

As was stated in § 1.2, our theory deals implicitly with monochromatic beams but we may estimate the *chromatic aberrations* by considering small variations of the beam energy. The most important contributions to the chromatic aberrations are the *paraxial chromatic aberrations* due to the variation of the function $m^{(2)}$ with beam energy, i.e. with Φ. We may make a qualitative appraisal of these aberrations by considering the equations (1.4.28) and (1.5.13). If Φ is increased, the coefficient of w in (1.4.28) and the coefficient

‡ Strictly speaking, all aberrations which we calculate are 'geometrical' aberrations as opposed to 'diffraction' aberrations, but the term will be used in a restricted sense to denote the aberrations due to terms higher than $m^{(2)}$ in the polynomial expansion of m.

of v in (1.5.13) are decreased so that the curvature of a ray with a given ordinate w or v is decreased. Since the curvature is always directed towards the axis, this means that the lens becomes less strongly convergent and that the image of an object therefore moves away from the lens.‡

These and other aberrations (the 'relativistic aberrations' and the aberrations due to asymmetries) will be considered in detail in Chapter 4. We must therefore bear in mind in the next two chapters the need to establish a method suitable for their calculation. Since optical instruments form acceptable images only if the aberrations are small, we may expect the calculation of aberrations to reduce essentially to the calculation of the change in image-forming properties of an optical system due to a small change in its constitution.

‡ This argument is due to O. Scherzer, ref. (27).

CHAPTER 2

CLASSICAL GEOMETRICAL OPTICS

2.1. Introduction

Although it would be possible to continue our discussion of rotationally symmetrical systems, applying well-known mathematical methods to the solution of the problems listed at the end of the last section,‡ it will be profitable at this stage to review the principal concepts of classical geometrical optics. In this way we shall establish such useful and basic rules as the 'imaging relations' and bring to light optical methods which may be applied in investigations of electron-optical systems.

The generic problem of geometrical optics is the solution of the ray equation. Hamilton's great contribution was to show that, since the ray equation is derived from a variational equation, its solution may be represented by a *characteristic function* whose arguments are the co-ordinates of an arbitrary pair of points and whose value is the *optical distance* between these points, i.e. the line integral weighted by the refractive index taken along the ray (characteristic) connecting the points (cf. eq. (2.2.11)).

This discovery alone does not solve the problem, for it remains to show how the characteristic function may be calculated. The methods which are most useful in light optics§ are based on the assumption that the refractive index is homogeneous between given surfaces and so cannot be applied to electron optics. The general method proposed by Hamilton is impracticable, for it requires the solution of a pair of simultaneous partial differential equations (eqs. (2.2.18)). However, it was shown by Jacobi and Liouville that the rays of an optical system may be found once one has obtained any complete solution of only one partial differential equation (eq. (2.7.5)), known as the *Hamilton-Jacobi equation*, which is formally identical with either of Hamilton's pair of equations.

‡ For this approach, see O. Scherzer, *Beiträge zur Elektronenoptik*, ed. H. Busch and E. Brüche (Barth, Leipzig, 1937), pp. 33–41.

§ See, for instance, ref. (10), pp. 36 and 57, or ref. (11), pp. 173 et seq.

Although the Hamilton-Jacobi equation is of great theoretical interest and probably could often be applied as an advantageous alternative to the ray equation,‡ there is another method for the calculation of the characteristic function which, although it is restricted to a certain class of problems, is much better suited to the needs of electron optics. As was indicated at the end of the last section, one often wishes to estimate in what way an electron-optical system, whose properties are known, is affected by slight changes in its constitution. It is therefore of great value to have a method for calculating that part of a characteristic function which represents a *perturbation* of an optical system. The method of *perturbation characteristic functions*, as it is here called, was introduced into light optics by Schwarzschild§ and into electron optics by Glaser.||

The order in which these ideas will be discussed in this chapter is as follows: We shall first deduce from the variational equation the general ray equation and the existence of the characteristic function. A later section will deal with perturbation characteristic functions. We shall then proceed to discuss the imaging relations, which one may regard as structural properties of image-forming ray assemblies. The imaging relations will be derived from the Lagrange invariant whose existence represents that element of the structure of an assembly of rays which is due to their optical nature. The last two sections deal with 'normal congruences', i.e. assemblies of rays which are—in a certain sense—optically indistinguishable from the type of assembly one obtains from a point source, and with the Hamilton-Jacobi equation. It is interesting to notice in the last section the close relationship between perturbation characteristic functions and the Hamilton-Jacobi equation.

2.2. Hamilton's characteristic functions

Let us denote by S the integral

$$S=\int_A^B n\,\mathrm{d}s \qquad (2.2.1)$$

‡ See refs. (20) and (21).

§ See, for instance, M. Born, *Optik* (Springer, Berlin, 1933), pp. 68–70.

|| W. Glaser, *Ann. Phys., Lpz.*, **18** (1933), 557–85; *Z. Phys.* **104** (1936), 157–60.

evaluated along an arbitrary path between the points A and B. The co-ordinates of points of the path may be expressed as functions of the arc length s, i.e. as $\mathbf{x}(s)$.

Now let the path of integration be varied so that the point $\mathbf{x}(s)$ is displaced to $\mathbf{x}(s)+\delta\mathbf{x}(s)$. Then, using the relations $ds^2=d\mathbf{x}^2$ and $\mathbf{l}=d\mathbf{x}/ds$, we find the following expressions for the increments of arc length and of the direction vector:

$$\frac{d\delta s}{ds}=\mathbf{l}.\frac{d\delta\mathbf{x}}{ds},\quad \delta\mathbf{l}=\frac{d\delta\mathbf{x}}{ds}-\left(\mathbf{l}.\frac{d\delta\mathbf{x}}{ds}\right)\mathbf{l}. \qquad (2.2.2)$$

Moreover, the refractive index, which was n at a point of the original path, becomes $n+\delta n$ at the corresponding point of the new path, where

$$\delta n=\delta\mathbf{x}.\frac{\partial n}{\partial\mathbf{x}}+\delta\mathbf{l}.\frac{\partial n}{\partial\mathbf{l}}. \qquad (2.2.3)$$

Since the variation in S is given by

$$\delta S=\int_A^B\left\{\delta n+n\frac{d\delta s}{ds}\right\}ds, \qquad (2.2.4)$$

we may, by means of (2.2.2) and (2.2.3), relate δS to the function $\delta\mathbf{x}(s)$. We find that

$$\delta S=\int_A^B\left\{\mathbf{p}.\frac{d\delta\mathbf{x}}{ds}+\delta\mathbf{x}.\frac{\partial n}{\partial\mathbf{x}}\right\}ds, \qquad (2.2.5)$$

where

$$\mathbf{p}=n\mathbf{l}+\frac{\partial n}{\partial\mathbf{l}}-\left(\mathbf{l}.\frac{\partial n}{\partial\mathbf{l}}\right)\mathbf{l}. \qquad (2.2.6)$$

The vector $\mathbf{p}$ defined by the formula (2.2.6) is of great importance when the vector $\mathbf{l}$ refers to the direction of a ray; $\mathbf{p}$ will then be termed the *ray vector*. We may note the interesting relation

$$\mathbf{p}.\mathbf{l}=n, \qquad (2.2.7)$$

i.e. that the projection of the ray vector upon the tangent to the ray is equal to the refractive index. It is found from (1.3.10) and (2.2.6) that the ray vector is given, in electron optics, by the formula

$$\mathbf{p}=p\mathbf{l}-\mathbf{A}, \qquad (2.2.8)$$

from which it is apparent that *the 'ray vector' of electron optics is identical with the canonical momentum of dynamics.*‡ One should

‡ It is important to distinguish between the scalar momentum p defined by (1.2.4) and components of the ray vector or canonical momentum $\mathbf{p}$; the latter will always carry suffixes such as p_r, p_i, or p_x, p_y, p_z which enumerate the components.

also note from (2.2.8) that the ray vector is parallel to the ray tangent only if the refractive index is locally isotropic.

On performing an integration by parts, we may put the equation (2.2.5) into the form

$$\delta S = \mathbf{p}_b . \delta\mathbf{x}_b - \mathbf{p}_a . \delta\mathbf{x}_a - \int_A^B \delta\mathbf{x} . \left\{ \frac{d\mathbf{p}}{ds} - \frac{\partial n}{\partial \mathbf{x}} \right\} ds \qquad (2.2.9)$$

which, it should be remembered, is valid for an arbitrary variation of an arbitrary path of integration. If we now specify that this path is to be a *ray*, it follows from the variational equation (1.3.9) that δS is to vanish for all $\delta\mathbf{x}(s)$ provided that $\delta\mathbf{x}_a = \delta\mathbf{x}_b = 0$. Hence the path of integration is a ray if and only if

$$\frac{d\mathbf{p}}{ds} = \frac{\partial n}{\partial \mathbf{x}}, \qquad (2.2.10)$$

which will be called the *ray equation*, is satisfied. The reader may verify that for purely electric fields, (2.2.10) may be put into the form (1.4.8).

On returning to (2.2.9) we see that, if the original path of integration is a ray and if $\delta\mathbf{x}_a = \delta\mathbf{x}_b = 0$, then $\delta S = 0$. This means that *if integrals* (2.2.1) *are evaluated along a ray and along a neighbouring path with the same terminal points, the difference contains no contribution which varies linearly with the displacement between the paths.*

Let us now denote by V the value of the integral (2.2.1) taken along the ray between the points A and B:

$$V = \int_A^B n\, ds. \qquad (2.2.11)$$

It is clear that V is defined only if such a ray exists; on the other hand, it is possible for two or more rays to pass through a given pair of points in which case V may be multiply valued. The latter case will be excluded from our considerations.

If the path of integration is varied so as to coincide with a neighbouring ray, the variation in V may be evaluated from (2.2.9) which becomes

$$\delta V = \mathbf{p}_b . \delta\mathbf{x}_b - \mathbf{p}_a . \delta\mathbf{x}_a. \qquad (2.2.12)$$

This is *Hamilton's differential relation* or 'equation of the characteristic function'; we shall see in § 2.6 that (2.2.12) may be interpreted as a mathematical formulation of Huyghens's principle. It is seen that V, whose value—as was stated in § 2.1—is the 'optical distance' between A and B, may be expressed as $V(\mathbf{x}_a, \mathbf{x}_b)$, i.e. as a function of the co-ordinates of the terminal points of the ray of integration; so expressed, it is Hamilton's 'characteristic function' or, more specifically, Hamilton's *point characteristic function*. In saying that this function 'characterizes' the optical coupling of the A- and B-spaces—i.e. the neighbourhoods of A and B—we imply that instruments are indistinguishable if their characteristic functions for the coupling between the object and image spaces are identical.‡ The existence of the point characteristic function

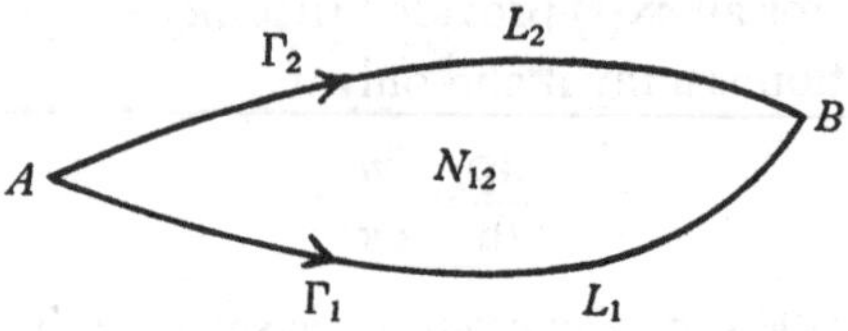

Fig. 2.1. Optical distance in a magnetic field.

together with the relation (2.2.12) must represent a complete statement of the laws of geometrical optics since the Hamilton-Jacobi equation, which will be derived from these premises, will be seen in § 2.7 to lead back to the variational equation with which we began.

Let us now suppose that A and B are *conjugate* so that a congruence—i.e. a two-parameter family of rays—passes through both A and B. We may restrict δ to refer to a displacement from a given ray of the assembly to a neighbouring ray; then $\delta\mathbf{x}_a = \delta\mathbf{x}_b = 0$ so that, from (2.2.12), $\delta V = 0$. It follows that V has the same value for all rays of the manifold; hence *the optical distance along all rays connecting a pair of conjugate points is the same.*

As a simple example, let us suppose that A and B are connected by an assembly of rays in a purely magnetic field. Let two of the rays Γ_1 and Γ_2 have lengths L_1 and L_2, respectively (fig. 2.1). We then find, by substituting (1.3.10) in (2.2.11), that

$$\int_{\Gamma_1} (p - \mathbf{1}.\mathbf{A})\,\mathrm{d}s = \int_{\Gamma_2} (p - \mathbf{1}.\mathbf{A})\,\mathrm{d}s \qquad (2.2.13)$$

which, since p is constant, may be rewritten as

$$p(L_2 - L_1) = \int_{\Gamma_1} \mathbf{A}.\mathrm{d}\mathbf{s} - \int_{\Gamma_2} \mathbf{A}.\mathrm{d}\mathbf{s}. \qquad (2.2.14)$$

‡ The couplings are indistinguishable also if the characteristic functions differ only by a constant.

The integrals on the right-hand side may be combined into an integral round a closed contour. Application of Stokes's theorem then gives

$$p(L_2-L_1)=N_{12}, \tag{2.2.15}$$

where N_{12} denotes the magnetic flux embraced by the rays; the flux is measured so as to make positive the flux produced by a current flowing from A to B along Γ_2 and returning along Γ_1.

The result may be stated as follows: *in the absence of electric fields, the magnetic flux embraced by two rays connecting a pair of conjugate points is equal to the product of the beam momentum and the difference in length of the rays.*

Let us now suppose that the points A and B are non-conjugate, so that any point in the neighbourhood of A is connected by a ray to any point in the neighbourhood of B. δV is then defined for arbitrary $\delta \mathbf{x}_a$ and $\delta \mathbf{x}_b$ so that the derivatives of V exist and are related to the ray vectors at A and B by the equations

$$\mathbf{p}_a=-\frac{\partial V}{\partial \mathbf{x}_a}, \quad \mathbf{p}_b=\frac{\partial V}{\partial \mathbf{x}_b}. \tag{2.2.16}$$

It should be noted that the differential form of a relation, such as (2.2.12), has the advantage over the differentiated form, such as (2.2.16), in that the former holds for *any* ray assembly whereas the latter does not. We may also observe that, where the refractive index is isotropic, the ray direction is parallel to the gradient of the characteristic function since, from (2.2.8), the ray direction and ray vector are then parallel. Another trivial result is that if $\mathbf{A}$ is changed to $\mathbf{A}+\operatorname{grad}\chi$, V is changed to $V+\chi_a-\chi_b$.

Since $\mathbf{l}$ is a unit vector, it follows from (2.2.8) that

$$(\mathbf{p}+\mathbf{A})^2=p^2, \tag{2.2.17}$$

and on combining this equation with the equations (2.2.16) we obtain

$$\left\{\frac{\partial V}{\partial \mathbf{x}_a}-\mathbf{A}_a\right\}^2=p_a^2, \quad \left\{\frac{\partial V}{\partial \mathbf{x}_b}+\mathbf{A}_b\right\}^2=p_b^2. \tag{2.2.18}$$

These are *Hamilton's partial differential equations* which in principle offer a method for the calculation of the characteristic function but which, as was pointed out in § 2.1, are unsuitable for practical calculations.

2.3. The Lagrange invariant

It was observed in the previous section that the existence of the point characteristic function together with the relation (2.2.12) which it satisfies comprise a complete formulation of the laws of geometrical optics. The question suggests itself: 'Is it possible to reduce the formulation to a differential relation involving only the ray vector?' This is equivalent to asking whether Hamilton's relation, which states that the structure of a ray assembly is related in a certain simple way to a scalar spatial function, may be replaced by a formulation of the essentially 'optical' structure of ray assemblies which involves only the ray-vector distribution. This reduction is possible and it will now be carried out.

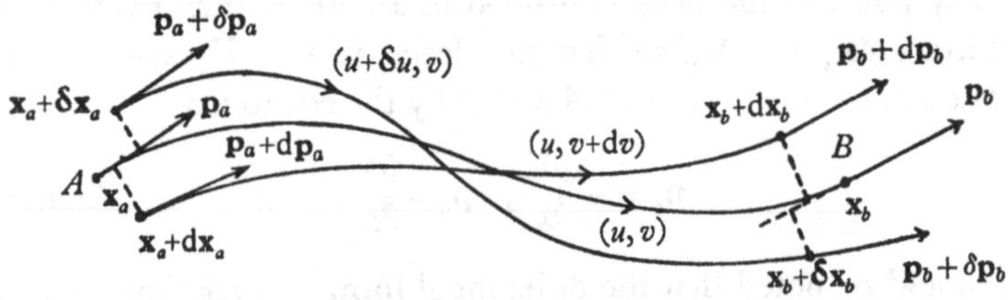

Fig. 2.2. Increments which make up the Lagrange invariant.

Let us suppose that the ray joining A and B may suffer separately or simultaneously two displacements; let the resulting increments in V and the co-ordinates be denoted by δV, etc., in one case and $\mathrm{d}V$, etc., in the other (fig. 2.2). The relation (2.2.12) holds for the former displacement by itself, but if this is followed by the second displacement the product of the two is represented by the relation

$$\mathrm{d}\delta V = \mathrm{d}\mathbf{p}_b . \delta\mathbf{x}_b + \mathbf{p}_b . \mathrm{d}\delta\mathbf{x}_b - \mathrm{d}\mathbf{p}_a . \delta\mathbf{x}_a - \mathbf{p}_a . \mathrm{d}\delta\mathbf{x}_a. \qquad (2.3.1)$$

Now the product must obviously be the same if the order of the displacements is reversed, i.e. $\mathrm{d}\delta V = \delta\mathrm{d}V$. Hence we find that

$$\delta\mathbf{p}_a . \mathrm{d}\mathbf{x}_a - \mathrm{d}\mathbf{p}_a . \delta\mathbf{x}_a = \delta\mathbf{p}_b . \mathrm{d}\mathbf{x}_b - \mathrm{d}\mathbf{p}_b . \delta\mathbf{x}_b, \qquad (2.3.2)$$

so that *the conbination* $\delta\mathbf{p} . \mathrm{d}\mathbf{x} - \mathrm{d}\mathbf{p} . \delta\mathbf{x}$ *of the differences in the position and ray vectors between three neighbouring rays whose typical points are* $\mathbf{x}(s)$, $\mathbf{x}(s)+\delta\mathbf{x}(s)$ *and* $\mathbf{x}(s)+\mathrm{d}\mathbf{x}(s)$ *is constant along the length of the rays.* This is the *Lagrange differential invariant.*

Let us suppose that the rays belong to a congruence enumerated by the parameters u and v which take the values (u, v), $(u+\delta u, v)$

and $(u, v+\mathrm{d}v)$ for the rays under consideration. In order to specify the terminal points uniquely, we must fix the values δs_a, etc., of the projections of the vector joining $\mathbf{x}_a$ to $\mathbf{x}_a+\delta\mathbf{x}_a$, etc., upon the original ray. Since the right-hand side of (2.3.2) cannot depend upon δs_a and $\mathrm{d}s_a$, neither can the left-hand side; similarly the right-hand side cannot depend upon δs_b and $\mathrm{d}s_b$. If we introduce the notation

$$\{u, v\} = \frac{\partial \mathbf{p}}{\partial v}\cdot\frac{\partial \mathbf{x}}{\partial u} - \frac{\partial \mathbf{p}}{\partial u}\cdot\frac{\partial \mathbf{x}}{\partial v}, \tag{2.3.3}$$

the Lagrange differential invariant is clearly equal to $\{u, v\}\,\delta u\,\mathrm{d}v$, so that the expression (2.3.3), which is known as the *Lagrange bracket*, is constant along any ray. For a triple manifold, i.e. a three-parameter family of rays, there will be three Lagrange brackets and, for a quadruple-manifold, six.

The relation (2.3.2) has been established in order that we may later derive from it the imaging relations. It therefore remains only to demonstrate that this relation will lead back to, and is therefore equivalent to, Hamilton's relation (2.2.12) from which it was derived.

It is well known that the necessary and sufficient condition for $\mathbf{F}.\mathrm{d}\mathbf{x}$ to be a complete differential, where $\mathbf{F}$ is a vector function of $\mathbf{x}$, is that curl $\mathbf{F}$ should vanish for all $\mathbf{x}$. If we take the scalar product of curl $\mathbf{F}$ with the element of area formed by the parallelogram with sides $\delta\mathbf{x}$ and $\mathrm{d}\mathbf{x}$, the condition becomes

$$\delta\mathbf{F}.\mathrm{d}\mathbf{x} - \mathrm{d}\mathbf{F}.\delta\mathbf{x} = 0 \tag{2.3.4}$$

for all $\delta\mathbf{x}$ and $\mathrm{d}\mathbf{x}$ and for all $\mathbf{x}$. The condition is the better represented by (2.3.4), since it may be applied even if $\mathbf{F}$ is not defined for all $\mathbf{x}$. If, for instance, we suppose $\mathbf{F}$ to be defined only on a surface, (2.3.4) implies only that the normal component of curl $\mathbf{F}$ vanishes.‡

If we now rearrange (2.3.2) as

$$(\delta\mathbf{p}_b.\mathrm{d}\mathbf{x}_b - \delta\mathbf{p}_a.\mathrm{d}\mathbf{x}_a) - (\mathrm{d}\mathbf{p}_b.\delta\mathbf{x}_b - \mathrm{d}\mathbf{p}_a.\delta\mathbf{x}_a) = 0, \tag{2.3.5}$$

we see, on adopting for the moment a six-dimensional space with co-ordinates $(x_{1a}, x_{2a}, x_{3a}, x_{1b}, x_{2b}, x_{3b})$ and considering the vector $(-p_{1a}, -p_{2a}, -p_{3a}, p_{1b}, p_{2b}, p_{3b})$, that (2.3.2) is the condition for $\mathbf{p}_b.\mathrm{d}\mathbf{x}_b - \mathbf{p}_a.\mathrm{d}\mathbf{x}_a$ to be a complete differential, i.e. for the existence of $V(\mathbf{x}_a, \mathbf{x}_b)$ satisfying (2.2.12).

‡ Cf. the discussion on p. 10.

By restricting our attention to a congruence, we may obtain an integrated representation of the relation (2.3.2). By retracing the step by which we arrived at (2.3.4), we see that (2.3.2) may be written as

$$\delta d\mathbf{S}_a \,.\, \operatorname{curl} \mathbf{p}_a = \delta d\mathbf{S}_b \,.\, \operatorname{curl} \mathbf{p}_b, \tag{2.3.6}$$

where $\delta d\mathbf{S} = \delta\mathbf{x} \wedge d\mathbf{x}$. On integrating the expressions in (2.3.6) by means of Stokes's theorem, we obtain

$$\int_{\Gamma_a} \mathbf{p} \,.\, d\mathbf{s} = \int_{\Gamma_b} \mathbf{p} \,.\, d\mathbf{s}, \tag{2.3.7}$$

where Γ_a and Γ_b now represent any two closed curves drawn round the same tube of rays (fig. 2.3).

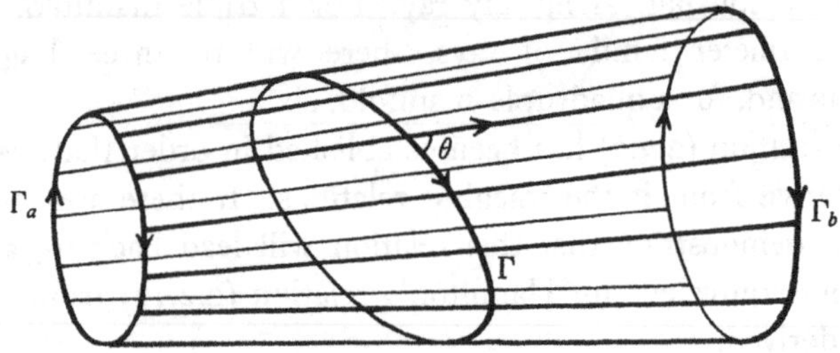

Fig. 2.3. The Poincaré invariant.

The relation (2.3.7) may be stated as follows: *the integral* $\oint \mathbf{p} \,.\, d\mathbf{s}$ *taken round any closed curve encircling a given tube of rays is a constant for that tube.* It is known as the *Poincaré invariant*, and it is easy to see from (2.2.8) that it may be written as

$$\int_{\Gamma} p \cos\theta \, ds - N_{\Gamma}, \tag{2.3.8}$$

where θ is the angle between the ray direction and the path of integration and N_Γ is the magnetic flux enclosed by the path of integration (fig. 2.3). It should be noticed that if a tube of rays emanates from a point, the Poincaré invariant vanishes, and hence that a tube of rays can be focused to a point only if the Poincaré invariant vanishes. Similar significance may be attached to the vanishing of the Lagrange invariant, but this will be discussed in §2.6.

Provided we distinguish the respective usages of the variable θ, it is not difficult to show that the equation (1.5.4) expresses the invariance of the function (2.3.8).

In the same way as it was shown that (2.3.2) is the condition for the existence of Hamilton's point characteristic function, it may be shown by other rearrangements that (2.3.2) is the condition for the existence of three other characteristic functions which were also discovered by Hamilton. These are the two 'mixed characteristic functions' $W(\mathbf{x}_a, \mathbf{p}_b)$ and $W^*(\mathbf{p}_a, \mathbf{x}_b)$ and the 'angle characteristic function' $T(\mathbf{p}_a, \mathbf{p}_b)$ which are defined by

$$W = V - \mathbf{p}_b \cdot \mathbf{x}_b, \tag{2.3.9}$$

$$W^* = V + \mathbf{p}_a \cdot \mathbf{x}_a \tag{2.3.10}$$

and

$$T = V + \mathbf{p}_a \cdot \mathbf{x}_a - \mathbf{p}_b \cdot \mathbf{x}_b, \tag{2.3.11}$$

and satisfy the differential relations

$$\delta W = -\mathbf{p}_a \cdot \delta \mathbf{x}_a - \mathbf{x}_b \cdot \delta \mathbf{p}_b, \tag{2.3.12}$$

$$\delta W^* = \mathbf{x}_a \cdot \delta \mathbf{p}_a - \mathbf{p}_b \cdot \delta \mathbf{x}_b \tag{2.3.13}$$

and

$$\delta T = \mathbf{x}_a \cdot \delta \mathbf{p}_a - \mathbf{x}_b \cdot \delta \mathbf{p}_b. \tag{2.3.14}$$

Let us consider the function $W(\mathbf{x}_a \cdot \mathbf{p}_b)$ and suppose that the field and the concomitant optical coupling are such that $\delta \mathbf{x}_a$ and $\delta \mathbf{p}_b$ may be prescribed arbitrarily. Then it follows from (2.3.12) that the derivatives of W exist and satisfy the equations

$$\mathbf{p}_a = -\frac{\partial W}{\partial \mathbf{x}_a}, \quad \mathbf{x}_b = -\frac{\partial W}{\partial \mathbf{p}_b}. \tag{2.3.15}$$

However, let us now suppose that B is in field-free space so that $\mathbf{p}_b = p_b \mathbf{l}_b$ and $\delta p_b = 0$ for any $\delta \mathbf{x}_b$. Since $\mathbf{l} . \delta \mathbf{l} = 0$, possible increments of $\mathbf{p}_b$ are now restricted by the relation

$$\mathbf{l}_b . \delta \mathbf{p}_b = 0, \tag{2.3.16}$$

so that the second of equations (2.3.15) breaks down.

If we adopt a co-ordinate frame (x_i, z) in the B-space, so chosen that $l_{zb} \neq 0$, (2.3.16) makes it possible to rewrite (2.3.12) as

$$\delta W = -\mathbf{p}_a \cdot \delta \mathbf{x}_a - \{x_{ib} - (z_b/l_{zb})\, l_{ib}\}\, \delta p_{ib}. \tag{2.3.17}$$

Hence if B is in field-free space W may be redefined as $W(\mathbf{x}_a, p_{ib})$ so that the number of its arguments is reduced from six to five. We then see from (2.3.17) that

$$\mathbf{p}_a = -\frac{\partial W}{\partial \mathbf{x}_a}, \quad x_{ib} = -\frac{\partial W}{\partial p_{ib}} + z_b(l_{ib}/l_{zb}). \tag{2.3.18}$$

Since, in light optics, the object and image are always located in isotropic homogeneous media, the number of arguments of the mixed characteristic functions may be reduced to five and that of the angle characteristic function to four. The same simplification does not invariably arise in electron optics, but wherever it does the mixed and angle characteristic functions are easier to handle than the point characteristic function.

In the remainder of this tract we shall restrict our attention to the point characteristic function. Of the other characteristic functions, that which we have written as $W(\mathbf{x}_a, p_{ib})$ is the most important; it is often employed by Glaser, to whose work the interested reader is referred.‡

2.4. Perturbation characteristic functions

It was observed in the introductory section of this chapter that in order to be able to calculate the aberrations of electron lenses, which were mentioned at the end of § 1.5, we should establish a method for calculating the change in the optical properties of an electron-optical system due to a perturbation of the system. In this section it will be shown that the first-order effects of perturbations of optical systems may be characterized by functions analogous to those introduced in the last two sections. These functions, which it is natural to name 'perturbation characteristic functions', will be derived from a differential relation similar in form to Hamilton's relation.

Let us suppose that the electron-optical system defined by (1.3.9) is subjected to a perturbation which so depends on a *perturbation parameter* ϖ that the system is in its unperturbed state when $\varpi=0$. Since the unperturbed system is described by its distribution of refractive index $n(\mathbf{x}, \mathbf{l})$, the perturbed system is described by the perturbed form of this distribution which we shall write as $\mathsf{P}n(\mathbf{x}, \mathbf{l})$. The latter may be expanded as

$$\mathsf{P}n = n + \varpi n^{\mathrm{I}} + \dots, \tag{2.4.1}$$

where n^{I} is the first derivative of $\mathsf{P}n$ with respect to ϖ, evaluated for $\varpi=0$, and so on. We may expand p and $\mathbf{A}$ in the same way, and it is readily found from (1.3.10) that

$$n^{\mathrm{I}} = p^{\mathrm{I}} - \mathbf{l}.\mathbf{A}^{\mathrm{I}}, \tag{2.4.2}$$

and from (1.2.4) that $$p^{\mathrm{I}} = p^{-1}(1+\phi)\,\phi^{\mathrm{I}}. \tag{2.4.3}$$

‡ W. Glaser, *Ann. Phys., Lpz.*, 4 (1949), 389–408.

The same notation may be applied to the perturbed forms of the position and ray vectors so that

$$\mathsf{P}\mathbf{x}=\mathbf{x}+\varpi\mathbf{x}^{\mathrm{I}}+\dots \qquad (2.4.4)$$

and

$$\mathsf{P}\mathbf{p}=\mathbf{p}+\varpi\mathbf{p}^{\mathrm{I}}+\dots, \qquad (2.4.5)$$

but since, for the time being, we shall not specify that the perturbed ray should have the same terminal points as the unperturbed ray, the perturbed form of the point characteristic function will be written as

$$\mathsf{P}V=V+\varpi{}^{*}V^{\mathrm{I}}+\dots. \qquad (2.4.6)$$

Since first-order perturbations may be combined linearly, we may regard ${}^{*}V^{\mathrm{I}}$ as the sum of two contributions, one due to the perturbation of the path in the unperturbed field and the other due to the perturbation of the field along the unperturbed path. On evaluating the former by means of (2.2.12)—or, rather, the equation derived from (2.2.12) by replacing V by S—we find that

$${}^{*}V^{\mathrm{I}}=\mathbf{p}_b.\mathbf{x}_b^{\mathrm{I}}-\mathbf{p}_a.\mathbf{x}_a^{\mathrm{I}}+V^{\mathrm{I}}, \qquad (2.4.7)$$

where‡

$$V^{\mathrm{I}}=\int_A^B n^{\mathrm{I}}\,\mathrm{d}s, \qquad (2.4.8)$$

the integral being evaluated along the *unperturbed* ray.

The first-order contribution to the perturbed form of (2.2.12) is found, from (2.4.4), (2.4.5) and (2.4.6), to be

$$\delta{}^{*}V^{\mathrm{I}}=(\mathbf{p}_b^{\mathrm{I}}.\delta\mathbf{x}_b+\mathbf{p}_b.\delta\mathbf{x}_b^{\mathrm{I}})-(\mathbf{p}_a^{\mathrm{I}}.\delta\mathbf{x}_a+\mathbf{p}_a.\delta\mathbf{x}_a^{\mathrm{I}}). \qquad (2.4.9)$$

so that, from (2.4.7), V^{I} satisfies the relation

$$\delta V^{\mathrm{I}}=(\mathbf{p}_b^{\mathrm{I}}.\delta\mathbf{x}_b-\mathbf{x}_b^{\mathrm{I}}.\delta\mathbf{p}_b)-(\mathbf{p}_a^{\mathrm{I}}.\delta\mathbf{x}_a-\mathbf{x}_a^{\mathrm{I}}.\delta\mathbf{p}_a). \qquad (2.4.10)$$

This relation summarizes the behaviour of the first-order perturbations of an optical system just as Hamilton's relation summarizes the behaviour of an unperturbed system. Equation (2.4.10) may therefore be called ***the first-order perturbation relation.***

It is interesting to notice that if we specify that $n^{\mathrm{I}}\equiv 0$ so that $V^{\mathrm{I}}\equiv 0$ also, and if we write $\mathbf{dx}$ and $\mathbf{dp}$ for $\varpi\mathbf{x}^{\mathrm{I}}$ and $\varpi\mathbf{p}^{\mathrm{I}}$, then (2.4.10) reduces to (2.3.2) so that the perturbation relation embraces the Lagrange-invariant relation.

So far we have placed no restrictions upon the ray perturbation except that it should be compatible with the perturbation of the

‡ We use the symbol V, although, since the terminal points are still arbitrary, this symbol is strictly inappropriate, in order to bring out the analogy between (2.4.10) and (2.2.12).

field, but it is possible to impose restrictions upon any two of the four variables $\mathbf{x}_a^{\mathrm{I}}$, $\mathbf{p}_a^{\mathrm{I}}$, $\mathbf{x}_b^{\mathrm{I}}$ and $\mathbf{p}_b^{\mathrm{I}}$, provided that increments of those variables may be prescribed arbitrarily. Let us therefore suppose that A and B are non-conjugate, so that $\delta\mathbf{x}_a$ and $\delta\mathbf{x}_b$ can be varied arbitrarily, and impose the condition

$$\mathbf{x}_a^{\mathrm{I}} = \mathbf{x}_b^{\mathrm{I}} = 0. \tag{2.4.11}$$

The relation (2.4.10) now reduces to

$$\delta V^{\mathrm{I}} = \mathbf{p}_b^{\mathrm{I}} . \delta\mathbf{x}_b - \mathbf{p}_a^{\mathrm{I}} . \delta\mathbf{x}_a, \tag{2.4.12}$$

so that V^{I} may be expressed as $V^{\mathrm{I}}(\mathbf{x}_a, \mathbf{x}_b)$; so expressed, it may be called *the first-order point perturbation characteristic function.* Since A and B are non-conjugate, it follows at once that

$$\mathbf{p}_a^{\mathrm{I}} = -\frac{\partial V^{\mathrm{I}}}{\partial \mathbf{x}_a}, \quad \mathbf{p}_b^{\mathrm{I}} = \frac{\partial V^{\mathrm{I}}}{\partial \mathbf{x}_b}. \tag{2.4.13}$$

The reader will have noticed the similarity between the relations (2.4.12) and (2.4.13) and the relations (2.2.12) and (2.2.16); the former are simply the perturbed forms of the latter. We might therefore expect that it is possible to characterize a perturbation by a function which is the perturbed form of any one of the other three characteristic functions W, W^* and T introduced in §2.3. It is easily verified that such functions, which would be written as W^{I}, $W^{*\mathrm{I}}$ and T^{I}, may be calculated from the same formula (2.4.8); they satisfy differential relations which are obtained from (2.4.10) by equating to zero $\mathbf{x}_a^{\mathrm{I}}, \mathbf{p}_b^{\mathrm{I}}$; $\mathbf{p}_a^{\mathrm{I}}, \mathbf{x}_b^{\mathrm{I}}$ or $\mathbf{p}_a^{\mathrm{I}}, \mathbf{p}_b^{\mathrm{I}}$: and they are similar in form to (2.3.12), (2.3.13) and (2.3.14).

However, it is also possible to derive from (2.4.10) two perturbation characteristic functions which have no counterparts in the study of unperturbed systems. If we put

$$\mathbf{x}_a^{\mathrm{I}} = \mathbf{p}_a^{\mathrm{I}} = 0 \tag{2.4.14}$$

or

$$\mathbf{x}_b^{\mathrm{I}} = \mathbf{p}_b^{\mathrm{I}} = 0, \tag{2.4.15}$$

we obtain either

$$\delta U^{\mathrm{I}} = \mathbf{p}_b^{\mathrm{I}} . \delta\mathbf{x}_b - \mathbf{x}_b^{\mathrm{I}} . \delta\mathbf{p}_b \tag{2.4.16}$$

or

$$\delta U^{*\mathrm{I}} = -\mathbf{p}_a^{\mathrm{I}} . \delta\mathbf{x}_a + \mathbf{x}_a^{\mathrm{I}} . \delta\mathbf{p}_a, \tag{2.4.17}$$

where U^{I} and $U^{*\mathrm{I}}$, which may be expressed as $U^{\mathrm{I}}(\mathbf{p}_b, \mathbf{x}_b)$ and $U^{*\mathrm{I}}(\mathbf{p}_a, \mathbf{x}_a)$, may also be calculated from the formula (2.4.8). We see from (2.4.16) that, if B is not in field-free space,

$$\mathbf{p}_b^{\mathrm{I}} = \frac{\partial U^{\mathrm{I}}}{\partial \mathbf{x}_b}, \quad \mathbf{x}_b^{\mathrm{I}} = -\frac{\partial U^{\mathrm{I}}}{\partial \mathbf{p}_b}; \tag{2.4.18}$$

similarly we deduce from (2.4.17) that if A is not in field-free space

$$\mathbf{p}_a^{\mathrm{I}} = -\frac{\partial U^{*\mathrm{I}}}{\partial \mathbf{x}_a}, \quad \mathbf{x}_a^{\mathrm{I}} = \frac{\partial U^{*\mathrm{I}}}{\partial \mathbf{p}_a}. \tag{2.4.19}$$

Now the quantities $\mathbf{p}_a^{\mathrm{I}}$, $\mathbf{x}_a^{\mathrm{I}}$ which appear in these equations denote the extent to which $\mathbf{p}_a$ and $\mathbf{x}_a$ must be displaced if the perturbation of the field is not to result in a displacement of $\mathbf{p}_b$ and $\mathbf{x}_b$. Hence the effect of the perturbation upon the ray is *effectively* to modify $\mathbf{p}_a$ and $\mathbf{x}_a$ by amounts which are the negatives of $\mathbf{p}_a^{\mathrm{I}}$ and $\mathbf{x}_a^{\mathrm{I}}$. Hence if we write

$$\mathbf{p}_a^{*\mathrm{I}} = \frac{\partial U^{*\mathrm{I}}}{\partial \mathbf{x}_a}, \quad \mathbf{x}_a^{*\mathrm{I}} = -\frac{\partial U^{*\mathrm{I}}}{\partial \mathbf{p}_a}, \tag{2.4.20}$$

the quantities on the left-hand sides of these formulae represent the effective displacement of the object point due to the field perturbation. This formulation would be useful, for instance, in determining how a perturbation would affect one's appraisal of the object being studied in an electron microscope.

Should A or B be in field-free space we may modify a perturbation characteristic function in the same way as we modified the mixed characteristic function W in §2.3. The functions U^{I} and $U^{*\mathrm{I}}$ are particularly interesting in that they relate the perturbations of the co-ordinates of a ray in either the A or B space to the unperturbed co-ordinates of the ray in the same space without explicit reference to the co-ordinates of the ray in the other space.

As a simple example of the application of perturbation characteristic functions, let us consider the behaviour of a purely magnetic system which undergoes a change of beam momentum. The *dispersion* of β-ray spectrometers and the chromatic aberration of lenses come under this heading.

Let us suppose that, under the perturbation, p is changed to $(1+\varpi)p$ so that $p^{\mathrm{I}}=p$. If we now divide the perturbed form of (1.3.8) by $(1+\varpi)$, the variational equation is unaffected, but the perturbation now appears as a perturbation of the magnetic field with $\mathbf{A}^{\mathrm{I}} = -\mathbf{A}$. It follows that the results to be obtained will apply for either an increase in momentum by the factor ϖ or a decrease in field strength by the same factor.

Let us consider an assembly of rays of which the typical member originates in a fixed point A and terminates in a variable point B. We shall adopt co-ordinates (x_i, z) in the B-space and consider only points B which lie in the plane β given by $z_b = 0$ (fig. 2.4). If we further assume that there is no normal component of field strength in this plane we may, by the results of § 1.3, so arrange the vector potential that $\mathbf{A}$ vanishes in β.

Provided that A is not focused in β, we may introduce the point perturbation characteristic function which, from (2.4.2) and (2.4.8), is given by

$$V^{\mathrm{I}} = pL, \tag{2.4.21}$$

where L is the arc length of the ray joining the points A and B. Clearly L may be expressed as $L(x_{ib})$, and we then have, in place of (2.4.13),

$$p^{\mathrm{I}}_{ib} = \frac{\partial V^{\mathrm{I}}}{\partial x_{ib}}. \tag{2.4.22}$$

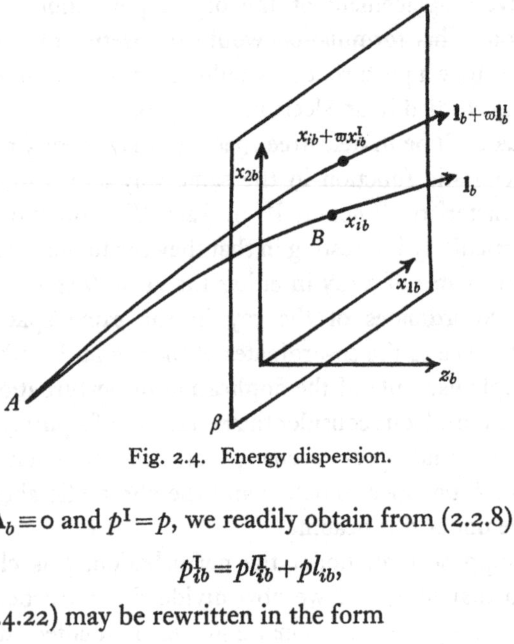

Fig. 2.4. Energy dispersion.

Since $\mathbf{A}_b \equiv 0$ and $p^{\mathrm{I}} = p$, we readily obtain from (2.2.8)

$$p^{\mathrm{I}}_{ib} = p l^{\mathrm{I}}_{ib} + p l_{ib}, \tag{2.4.23}$$

so that (2.4.22) may be rewritten in the form

$$l^{\mathrm{I}}_{ib} + l_{ib} = \frac{\partial L}{\partial x_{ib}}. \tag{2.4.24}$$

Equation (2.4.24) expresses the perturbation of the ray direction cosines in terms of certain derivatives of the arc length of the unperturbed rays.

We saw in the example that it was possible to transform the perturbation of the refractive index without modifying the results to which it led. We may easily obtain the general form of this transformation.

Since we do not affect the variational equation (1.3.9) by multiplying n by any quantity which is independent of $\mathbf{x}$ and $\mathbf{l}$, we may multiply the perturbed form of the refractive index (2.4.1) by any function of ϖ. Such a function may be expanded as

$$\pi(\varpi) = 1 + \lambda\varpi + O(\varpi^2), \tag{2.4.25}$$

provided that $\pi(0) = 1$, and we find that the new form of n^{I}, which we shall write as $^*n^{\mathrm{I}}$, is given by

$$^*n^{\mathrm{I}} = n^{\mathrm{I}} + \lambda n. \tag{2.4.26}$$

We may now state the rule: *it is possible to add any multiple of* n *to* n^{I} *without affecting the results of the perturbation calculations.*

Although it has been appropriate to introduce the perturbation parameters explicitly in this introductory exposition of perturbation characteristic functions, it is always possible to get rid of the parameters by writing Δf, etc., in place of ϖf^{I}, etc. Since we shall be dealing only with first-order perturbations, no confusion is likely to arise from this simplification, but it is safer to work in terms of perturbation parameters if higher-order effects are to be considered.‡

2.5. The reciprocal and imaging relations

It was shown in §2.3 that the laws of geometrical optics may be represented by a differential relation (2.3.2) which connects the configurations of three neighbouring rays at two distinct 'points' of their path. In this section we shall establish certain useful optical laws by applying the Lagrange relation to various special manifolds.§

The first reciprocal relation. Let us consider an assembly of rays which intersect a pair of planes α and β. Suppose that a typical ray meets these planes at the points A and B (fig. 2.5). We may choose two sets of co-ordinates (x_i, z) which have A and B as their origins and are so oriented that their z-axes are normal to α and β.

If A and B are non-conjugate, we may suppose the terminal

‡ See, for instance, the discussion of §5.3.

§ This section follows closely the corresponding sections of M. Herzberger's *Strahlenoptik* (Springer, Berlin, 1931).

points of the ray to be displaced first to the points $(\delta x_{ia}, 0)$ and $(0, 0)$ and then to the points $(0, 0)$ and $(dx_{ib}, 0)$. On noting that

$$dp_{ia} = p_a\, dl_{ia} \quad \text{and} \quad \delta p_{ib} = p_b\, \delta l_{ib},$$

we see that (2.3.2) reduces to

$$p_a\, dl_{ia}\, \delta x_{ia} + p_b\, \delta l_{ib}\, dx_{ib} = 0, \tag{2.5.1}$$

which we shall name *the first reciprocal relation.*

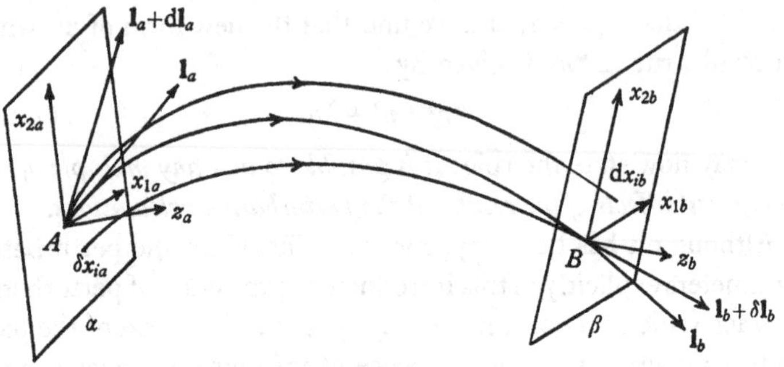

Fig. 2.5. The first reciprocal relation.

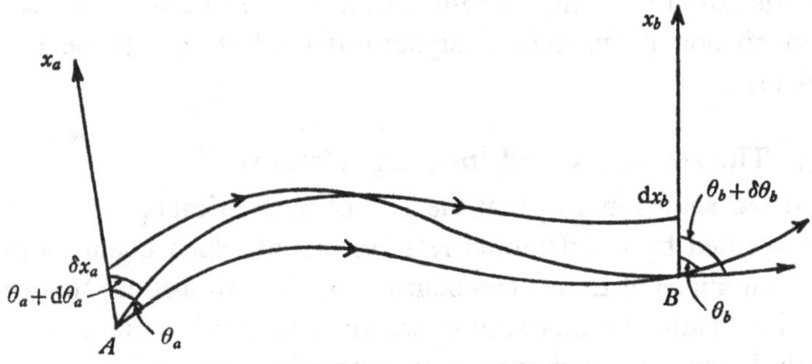

Fig. 2.6. The first reciprocal relation.

The same relation may be obtained in a different form by assuming that the terminal points may be displaced only along certain lines, say the x_1-axes. If $\delta x_{1a} = \delta x_a$, $\delta x_{2a} = 0$, $dx_{1b} = dx_b$, $dx_{2b} = 0$, $l_{1a} = \cos\theta_a$ and $l_{1b} = \cos\theta_b$ (fig. 2·6), (2.5.1) takes the form

$$p_a \sin\theta_a\, d\theta_a\, \delta x_a + p_b \sin\theta_b\, \delta\theta_b\, dx_b = 0. \tag{2.5.2}$$

It is implicit in (2.5.2) that if a single manifold connects each pair of points of the lines, θ_a and θ_b are constant for that manifold. A particular case of the relation (2.5.2) was first discovered by Huyghens.

The intensity of a beam. Let us suppose that the ray through A and B is, at these two points, normal to α and β, and let us denote by $\delta\mathrm{d}J$ the current which is carried by that part of the electron beam which passes through the rectangles whose vertices are (0, 0, 0), $(\delta x_{1a}, 0, 0)$, $(\delta x_{1a}, \delta x_{2a}, 0)$, $(0, \delta x_{2a}, 0)$ in α and (0, 0, 0), $(\mathrm{d}x_{1b}, 0, 0)$, $(\mathrm{d}x_{1b}, \mathrm{d}x_{2b}, 0)$, $(0, \mathrm{d}x_{2b}, 0)$ in β, whose areas we shall denote by δS_a and $\mathrm{d}S_b$. If the beam subtends solid angles $\mathrm{d}\Omega_a$ and $\delta\Omega_b$ at A and B, we may define the *current intensity* $\mathscr{I}$ at A and B by

$$\delta\mathrm{d}J = \mathscr{I}_a \delta S_a \mathrm{d}\Omega_a = \mathscr{I}_b \mathrm{d}S_b \delta\Omega_b. \tag{2.5.3}$$

By a suitable orientation of axes,‡ the relation between δl_{ib} and δx_{ia} may be put in the diagonal form

$$\delta l_{ib} = M_i \delta x_{ia} \quad (i \text{ n.t.b.s.}), \tag{2.5.4}$$

where the abbreviation 'n.t.b.s.' means 'not to be summed over'. We find, on substituting (2.5.4) in (2.5.1), that

$$\mathrm{d}l_{ia} = -(p_b/p_a)\, M_i \mathrm{d}x_{ib} \quad (i \text{ n.t.b.s.}). \tag{2.5.5}$$

On observing that $\delta S_a = \delta x_{1a}\delta x_{2a}$, $\mathrm{d}\Omega_a = \mathrm{d}l_{1a}\mathrm{d}l_{2a}$, etc., we may easily deduce from (2.5.4) and (2.5.5) that

$$p_a^2 \delta S_a \mathrm{d}\Omega_a = p_b^2 \mathrm{d}S_b \delta\Omega_b, \tag{2.5.6}$$

from which it follows at once that

$$\mathscr{I}_a/\mathscr{I}_b = p_a^2/p_b^2. \tag{2.5.7}$$

Hence *the current intensity of any narrow beam varies along its length as the square of the beam momentum.*

Since the energy carried by the current $\delta\mathrm{d}J$ at any point is $\delta\mathrm{d}J\,\phi$, *the energy intensity varies as* ϕp^2.

Semi-telescopic imaging. Let us suppose that all the rays leaving any point of α indefinitely close to A are parallel in the B-space which is field-free. Such imaging is termed 'semi-telescopic' for the element of α enclosing A would be imaged upon a plane placed 'at infinity' in the B-space.

Since x_{ia} determines l_{ib} uniquely, we may write

$$\delta l_{ib} = M_{ij} \delta x_{ja}. \tag{2.5.8}$$

‡ If $\delta l_{ib} = M_{ij}\delta x_{ja}$, the 'circle' $\delta l_{ib}\delta l_{ib} = \epsilon$ is the 'image' of the ellipse $M_{ij}M_{ik}\delta x_{ja}\delta x_{ka} = \epsilon$. If we adopt the principal axes of this ellipse as co-ordinate axes in the x_{ia} plane and the 'image' of these axes as the co-ordinate axes in the x_{ib} plane, the relation must reduce to the form (2.5.4).

It then follows from (2.5.1) that

$$\mathrm{d}l_{ja} = -(p_b/p_a)\,\mathrm{d}x_{ib} M_{ij}, \tag{2.5.9}$$

where d refers to the displacement of any ray through A to a neighbouring ray through A. Since M_{ij} is independent of x_{ib} and l_{ja}, (2.5.9) may be integrated to give

$$l_{ja} = -(p_b/p_a)\,x_{ib} M_{ij} + N_j. \tag{2.5.10}$$

Let us assume that imaging is such and the axes are so chosen that $M_{ij} = M\delta_{ij}$ and $x_{ib} = l_{ib} = 0$ when $x_{ia} = l_{ia} = 0$; systems of rotational symmetry satisfy these requirements (fig. 2.7). If we take $x_{1a} = r_a$, $x_{2a} = 0$, $x_{1b} = r_b$, $x_{2b} = 0$ and write $l_1 + \mathrm{i}l_2 = \sin\theta\,\mathrm{e}^{\mathrm{i}\psi}$, (2.5.8) becomes

$$\delta\theta_b = M\delta r_a, \quad \delta\psi_b = 0, \tag{2.5.11}$$

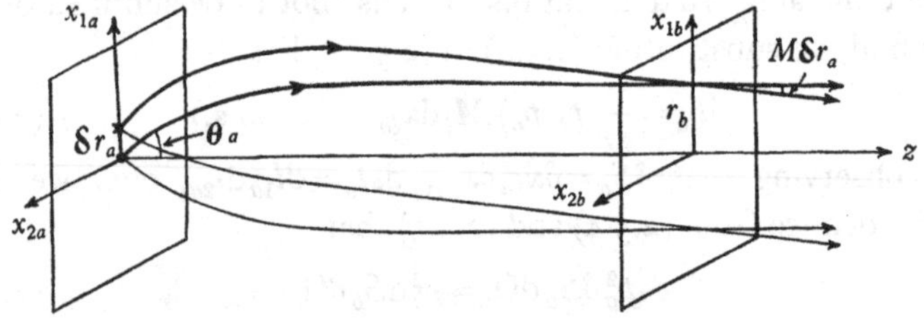

Fig. 2.7. Semi-telescopic imaging.

and the imaging relation (2.5.10) becomes

$$\sin\theta_a = -(p_b/p_a)\,Mr_b, \quad \psi_a = 0. \tag{2.5.12}$$

The relation (2.5.12) may be looked upon as the condition which must be satisfied by the rays passing through A in order that points in the neighbourhood of A may be imaged at infinity. For purely electric systems of rotational symmetry, (2.5.12) follows immediately from (2.5.2).

The second reciprocal relation. We shall again consider an assembly of rays meeting the planes α and β, but we shall now suppose that A and B are conjugate and that for any point of α close to A there is a conjugate point in β close to B (fig. 2.8). Adopting the same co-ordinates as before, we may write the relation between the permitted displacements of the terminal points as

$$\delta x_{ib} = M_{ij}\,\delta x_{ja}. \tag{2.5.13}$$

If we now assume d to refer to the displacement of any ray through A and B to a neighbouring ray with the same terminal points, we find from (2.3.2) that, since $d\mathbf{x}_a = d\mathbf{x}_b = 0$ and $\delta z_a = \delta z_b = 0$,

$$p_a\, dl_{ia}\, \delta x_{ia} = p_b\, dl_{ib}\, \delta x_{ib}. \tag{2.5.14}$$

This is ***the second reciprocal relation.***

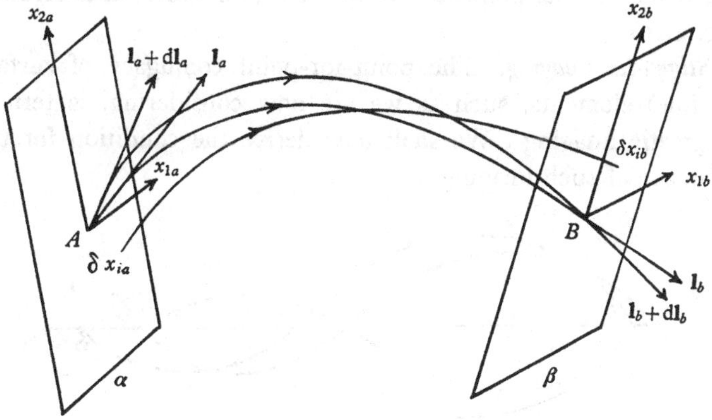

Fig. 2.8. The second reciprocal relation.

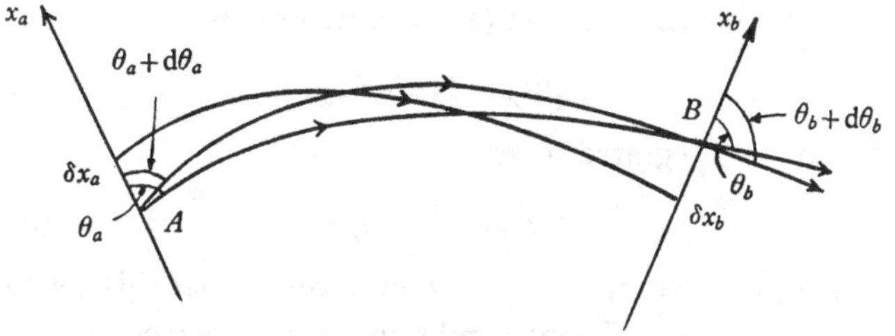

Fig. 2.9. The second reciprocal relation.

This relation may also be put into a simpler form if we consider conjugate sets of points distributed along two line elements rather than over two surface elements. If $\delta x_{1a} = \delta x_a$, $\delta x_{2a} = 0$, $\delta x_{1b} = \delta x_b$, $\delta x_{2b} = 0$, $l_{1a} = \cos\theta_a$ and $l_{1b} = \cos\theta_b$ (fig. 2.9), (2.5.14) becomes

$$p_a \sin\theta_a\, d\theta_a\, \delta x_a = p_b \sin\theta_b\, d\theta_b\, \delta x_b. \tag{2.5.15}$$

Special cases of this relation, such as (1.4.20), are associated with the names of Lagrange, Helmholtz and Clausius.

If we suppose that $\delta x_b = M\delta x_a$ in the above relation and that there exists a ray for which $\theta_a=\theta_b=0$, (2.5.15) gives upon integration the equation

$$p_a \sin^2 \tfrac{1}{2}\theta_a = M p_b \sin^2 \tfrac{1}{2}\theta_b. \tag{2.5.16}$$

If we interpret the x_a- and x_b-axes as portions of the axis of a rotationally symmetrical system in the object and image spaces, we recognize (2.5.16) to be *Herschel's condition*‡ which is to be satisfied if a range of axial points are to be free from spherical aberration (fig. 2.10).

Stigmatic imaging. The point-for-point conjugacy of surface (or line) elements, such as we are now considering, is termed 'stigmatic imaging'. We shall now derive the condition for the existence of such imaging.

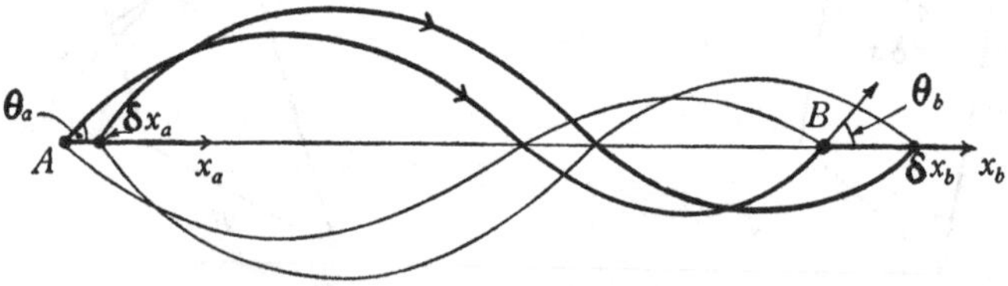

Fig. 2.10. Herschel's condition.

On combining (2.5.13) and (2.5.14), we obtain

$$p_a \, \mathrm{d}l_{ja} = p_b \, \mathrm{d}l_{ib} M_{ij}, \tag{2.5.17}$$

which may be integrated to give

$$l_{ja} = (p_b/p_a)\, l_{ib} M_{ij} + N_j. \tag{2.5.18}$$

By working back from (2.5.18) we may verify that this relation is the necessary and sufficient condition for the surface elements at A and B to be stigmatically imaged according to (2.5.13).

Let us suppose that the ray through A and B which is tangential to the z_a-axis at A is also tangential to the z_b-axis at B, and that we may take $M_{ij}=M\delta_{ij}$. These requirements are satisfied if a surface element normal to the axis of a system of rotational symmetry is imaged exactly (fig. 2.11). We may again write $x_{1a}=r_a$, $x_{2a}=0$, $x_{1b}=r_b$, $x_{2b}=0$ and $l_1+\mathrm{i}l_2=\sin\theta\, \mathrm{e}^{\mathrm{i}\psi}$, whereupon (2.5.18) takes the form

$$p_a \sin\theta_a = M p_b \sin\theta_b, \tag{2.5.19}$$

‡ See, for instance, H. H. Hopkins, *Wave Theory of Aberrations* (Oxford University Press, 1950), p. 45.

together with the trivial condition $\psi_a = \psi_b$. Equation (2.5.19) is the Abbe-Helmholtz *sine condition*. It clearly represents the condition that the aberration which varies with the first power of the object co-ordinates, i.e. the *coma*, should vanish.‡

Telescopic imaging. It is easy to see that the relation (2.5.14) would again be obtained if we specified that dl_{ia} and dl_{ib} satisfied a linear relationship independent of x_{ia} and x_{ib}. Since part of the plane at infinity in the A-space is then imaged upon the plane at infinity in the B-space, if these spaces are field-free, such imaging is termed 'telescopic'. If we start with

$$dl_{ib} = M_{ij}\, dl_{ja} \qquad (2.5.20)$$

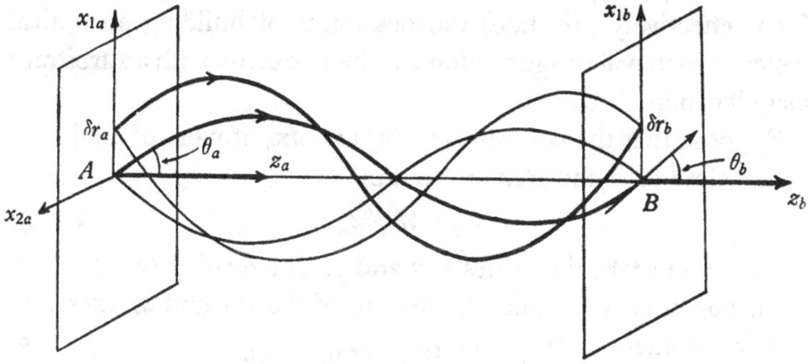

Fig. 2.11. The sine condition.

instead of with (2.5.13), we obtain in place of (2.5.18)

$$x_{ja} = (p_b/p_a)\, x_{ib} M_{ij} + N_j. \qquad (2.5.21)$$

Moreover, if we assume the system to be of rotational symmetry, we obtain in place of (2.5.19) the condition

$$r_a = (p_b/p_a)\, M r_b. \qquad (2.5.22)$$

The paraxial magnification relation. Let us suppose that two surface elements normal to the axis of a system of rotational symmetry are imaged with (transverse) magnification M_t, and that, in the same neighbourhood, elements of the axis are imaged with (longitudinal) magnification M_l. Then we see from (2.5.16) and (2.5.19) that

$$p_a \sin^2 \tfrac{1}{2}\theta_a = M_l\, p_b \sin^2 \tfrac{1}{2}\theta_b \qquad (2.5.23)$$

and

$$p_a \sin \theta_a = M_t\, p_b \sin \theta_b. \qquad (2.5.24)$$

‡ See, for instance, Hopkins, op. cit. pp. 39 and 55.

These are compatible for all values of θ_a and θ_b only if

$$M_t = M_l = p_a/p_b, \tag{2.5.25}$$

but (2.5.23) and (2.5.24) are satisfied to the paraxial approximation if only

$$M_l = (p_b/p_a)\,M_t^2. \tag{2.5.26}$$

This is the 'paraxial magnification relation'.

The Bruns-Klein theorem. We shall now prove that the result represented by (2.5.25), and so far proved only for rotationally symmetrical systems, is generally true: *If two volume elements are imaged exactly, the magnification is isotropic and equal to the ratio of the beam momentum at the object point to that at the image point.* This is the principal 'impossibility' theorem of geometrical optics for it effectively precludes the possibility of building an optical system which will image a volume element exactly with an arbitrary magnification.

Suppose that the co-ordinates $\delta\mathbf{x}_a$ and $\delta\mathbf{x}_b$ of pairs of conjugate points close to A and B are related by

$$\delta x_{rb} = M_{rs}\,\delta x_{sa}, \tag{2.5.27}$$

where r and s take the values 1, 2 and 3. The relation (2.5.27) may be diagonalized by a suitable rotation of the $\mathbf{x}_a$- and $\mathbf{x}_b$-axes,‡ so that it becomes

$$\delta x_{rb} = M_r\,\delta x_{ra} \quad (r \text{ n.t.b.s.}). \tag{2.5.28}$$

If the symbol d again refers to the displacement of any ray through A and B to a neighbouring ray which also passes through A and B, we find from (2.5.28) and (2.3.2) that

$$\mathrm{d}l_{ra} = (p_b/p_a)\,M_r\,\mathrm{d}l_{rb} \quad (r \text{ n.t.b.s}), \tag{2.5.29}$$

or, upon integration, that

$$l_{ra} = (p_b/p_a)\,M_r\,l_{rb} + N_r \quad (r \text{ n.t.b.s.}). \tag{2.5.30}$$

Since $\mathbf{l}$ is a unit vector, $l_{ra}^2 = 1$ and $l_{rb}^2 = 1$, but it is easily verified that—ignoring the possibility that either p_b or M_r vanishes—these conditions are incompatible unless $N_r = 0$ and

$$M_r = \pm p_a/p_b; \tag{2.5.31}$$

this proves the theorem.

Bruns was the first to prove that the magnification is isotropic, but Klein was the first to give the value of the magnification.§

‡ See p. 41, n. ‡.

§ F. Klein's ingenious proof is reproduced in E. T. Whittaker's *Theory of Optical Instruments* (Camb. Math. Tracts, no. 7, 1907), p. 47.

Since we have defined the various types of imaging, and since Hamilton's four characteristic functions have been introduced in earlier sections, we should indicate in which cases the various characteristic functions are appropriate or inappropriate.

Stigmatic imaging entails a relation between the increments $\delta\mathbf{x}_a$ and $\delta\mathbf{x}_b$: semi-telescopic imaging between either $\delta\mathbf{x}_a$ and $\delta\mathbf{p}_b$ or $\delta\mathbf{p}_a$ and $\delta\mathbf{x}_b$, and telescopic imaging between $\delta\mathbf{p}_a$ and $\delta\mathbf{p}_b$. It follows at once that these four types of imaging preclude the use of $V(\mathbf{x}_a, \mathbf{x}_b)$, $W(\mathbf{x}_a, \mathbf{p}_b)$, $W^*(\mathbf{p}_a, \mathbf{x}_b)$ and $T(\mathbf{p}_a, \mathbf{p}_b)$, respectively. The appropriate characteristic functions would be W, V, T and W^*, for it is then possible to find $\mathbf{x}_b$ in terms of $\mathbf{x}_a$ and $\mathbf{p}_b$, $\mathbf{p}_b$ in terms of $\mathbf{x}_a$ and $\mathbf{x}_b$, $\mathbf{x}_b$ in terms of $\mathbf{p}_a$ and $\mathbf{p}_b$, and $\mathbf{p}_b$ in terms of $\mathbf{p}_a$ and $\mathbf{x}_b$, respectively.

2.6. Normal congruences

In discussing the properties of an optical instrument, it is necessary to consider a quadruple manifold, for a fourfold infinity of rays connects the object space to the image space. However, such a manifold may be regarded as a family of manifolds of lower degree which are consequently simpler to work with. The double manifold or *congruence* is particularly useful if we specify that one and only one ray of such a manifold passes through each point of space except that a single infinity of rays may pass through each point of certain singular curves and the complete double infinity may pass through certain singular points.

If a congruence is enumerated by parameters u and v, the Lagrange bracket $\{u, v\}$, defined by (2.3.3), is constant along each ray. Now if the parameters u, v are replaced by another set u', v', the new Lagrange bracket is given by

$$\{u', v'\} = \{u, v\}\, \partial(u, v)/\partial(u', v'), \tag{2.6.1}$$

where $\partial(u, v)/\partial(u', v')$ is the Jacobian determinant which is non-zero for every non-singular transformation. It follows that if the Lagrange invariant vanishes for one set of parameters, it vanishes for every set so that the vanishing of this invariant must have physical significance. In this section we shall investigate the consequences of the following definition: *if the Lagrange invariant vanishes for every ray of a congruence, the congruence forms a normal congruence.*

An immediate consequence of our definition is that a normal congruence remains a normal congruence in passing through any optical system whatever. The proof of this statement, which is known as the *Malus-Dupin theorem*, lies in the premises of our definition, i.e. in §2.3.

It is easy to see from (2.3.3) that *the congruence formed by the rays emanating from a point is a normal congruence*, for

$$\partial \mathbf{x}/\partial u = \partial \mathbf{x}/\partial v = 0$$

at that point. It follows at once that *a necessary condition that a congruence may be brought to a point focus is that it should be a normal congruence.* This corroborates the similar observation concerning the Poincaré invariant made in §2.3, for it is obvious that *the Poincaré invariant vanishes for every tube of rays formed from a normal congruence.*

Since the Poincaré integral, taken round any closed path, vanishes, the integral defined by

$$V = \int_0^P \mathbf{p} . d\mathbf{s} \tag{2.6.2}$$

must be independent of the path of integration between the points O and P; this is known as the *Hilbert integral.* If O is an arbitrarily chosen origin, V is a function only of the position co-ordinates of P and so may be expressed as $V(\mathbf{x})$. There seems to be no accepted name for $V(\mathbf{x})$, so it is proposed that we adopt the obvious term 'optical potential'; the relation between this function and Hamilton's point-characteristic function will soon become clear.

It is seen from (2.6.2) that, since the path of integration is arbitrary,

$$\mathbf{p} = \frac{\partial V}{\partial \mathbf{x}}, \tag{2.6.3}$$

which is identical with the second of equations (2.2.16). We now see that the ray-vector distribution of a normal congruence may be expressed as the gradient of a scalar distribution so that *the ray vectors are normal to the surfaces $V(\mathbf{x}) =$ constant.* This justifies the name 'normal congruence'. Another consequence of (2.6.3) is that

$$\operatorname{curl} \mathbf{p} = 0 \tag{2.6.4}$$

throughout the volume of the congruence.

Let us now consider the surface α defined by $V(\mathbf{x})=0$, which passes through O, and let us divide the path of integration of (2.6.2) into two parts: any curve lying in α which joins O to P^*, the point where the ray through P meets α, and the ray P^*P (fig. 2.12). Clearly the first part does not contribute to the integral so that $V(\mathbf{x})$ *is equal to* $V(\mathbf{x}^*, \mathbf{x})$, where the latter notation is preserved for the point-characteristic function of § 2.2.

A necessary and sufficient condition that a congruence should be normal is that the Lagrange invariant should vanish for some point of every ray. Let us denote by $\mathbf{x}^*(u, v)$ the points of intersection

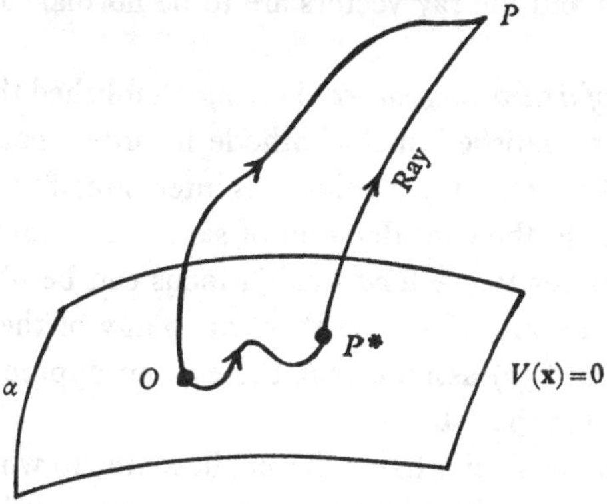

Fig. 2.12. The Hilbert integral.

of the rays (u, v) of the congruence with a surface α. Then we see from (2.3.3) that

$$\{u, v\} = -\mathbf{S}^*_{uv} . \operatorname{curl} \mathbf{p}^*, \tag{2.6.5}$$

where $\mathbf{S}^*_{uv} = (\partial \mathbf{x}^*/\partial u) \wedge (\partial \mathbf{x}^*/\partial v)$. Hence *a necessary and sufficient condition for the congruence to be normal is that the normal component of* curl $\mathbf{p}$ *should vanish over any surface intersecting the congruence.*

This condition may be put into another form by referring to the paragraph which contains (2.3.4). We then see that *a necessary and sufficient condition that a congruence should be normal is that* $\mathbf{p} . d\mathbf{s}$ *should be a complete differential in an arbitrary surface intersecting the rays.* This is in turn equivalent to the requirement that the Poincaré invariant should vanish for every contour taken in an arbitrary surface.

The cathode condition. If electrons leave a metallic cathode with negligible energy in the presence of an electric field, their initial motion is normal to the surface of the cathode so that the rays form a congruence. Since $p=0$ in the surface, we see from (2.3.8) that the Poincaré invariant of a tube of rays is equal to the magnetic flux embraced by the closed curve in which the tube intersects the cathode surface. It is now obvious that *electrons emitted from a cathode form a normal congruence if and only if the magnetic field has no component normal to the surface of the cathode.* It will be noted that this is precisely the condition, obtained in § 1.3, for the refractive index to be possibly isotropic over a surface. This is easily understood, for the latter expresses the condition which must be satisfied if both the rays and the ray vectors are to be normal to the cathode surface.

Best focus of a skew congruence. Having established the conditions which must be satisfied at the cathode in order that the emitted beam might be focused to a point, it is interesting for us to go back for a moment to the consideration of skew, i.e. non-normal, congruences in order to see how small a focus can be obtained if we (*a*) take into account the finite thermal energy of the electrons at the cathode and (*b*) assume that there is an appreciable normal magnetic field at the cathode.

Let us first consider what is the smallest area to which a tube of rays with a Poincaré invariant P can be reduced in field-free space. We may assume that the tube will be rotationally symmetrical; the rays will therefore form a set of generators of a hyperboloid of revolution (fig. 2.13). Let the minimum radius be r_f, the half-angle of the asymptotic cone θ_f, and the beam momentum p_f. Then we see that

$$P=2\pi p_f r_f \sin\theta_f. \tag{2.6.6}$$

(*a*) Let us now consider the electrons emitted at the cathode with momentum p_c upon the circumference of a circle of radius r_c. Of the tubes of rays which we can form from this triple manifold, that which has the largest Poincaré invariant is obviously that formed from rays whose initial direction lies along the circumference. For this tube

$$P=2\pi p_c r_c. \tag{2.6.7}$$

Upon combining (2.6.6) and (2.6.7), we see that the smallest focus which can be obtained from a beam originating in a cathode of radius r_c has the radius given by

$$r_f=p_c r_c/p_f \sin\theta_f. \tag{2.6.8}$$

(*b*) If we again ignore the initial energy of the electrons but now suppose that there exists a magnetic field of strength H_c normal to the cathode, we see that the Poincaré invariant for the rays originating on the circumference of the same circle as before is given by

$$P = \pi r_c^2 H_c. \tag{2.6.9}$$

We now obtain for the radius of the smallest focus which can be obtained from a cathode of radius r_c the formula

$$r_f = r_c^2 H_c / 2p_f \sin\theta_f. \tag{2.6.10}$$

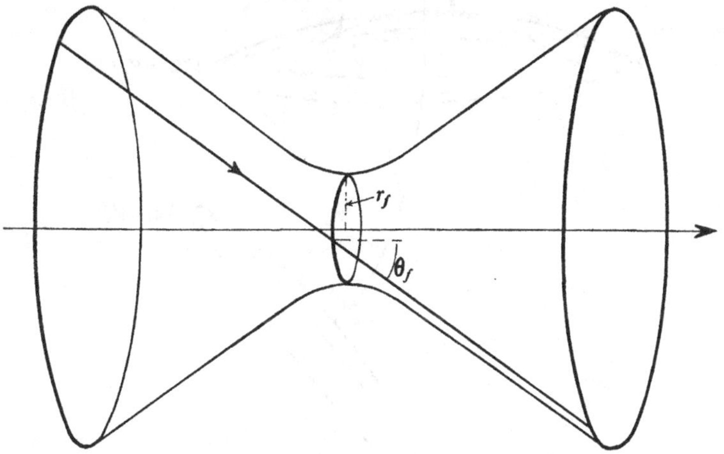

Fig. 2.13. Rotationally symmetrical skew congruence.

It is interesting to apply (2.6.8) and (2.6.10) to examples which might be met in electron microscopy. Consider electrons with a thermal energy of 0·1 volt which leave a cathode of diameter 0·1 mm. and are accelerated to 50,000 volts, at which energy they are brought to a focus in a cone with half-angle 0·0003 radian. On noting that $p_f/p_c \approx 700$, we see at once from (2.6.8) that the diameter of the beam at best focus is about 0·5 mm. If there is a stray magnetic field of strength 1 gauss normal to the cathode, $H_c \approx 0{\cdot}0006$ and $p_f \approx 0{\cdot}45$ in electron-optical units, so that (2.6.10) leads to an estimate of about 0·001 mm. for the minimum diameter of the beam. Clearly the effect of thermal energies outweighs the effect of the stray field.

If we are given a function $V^*(\mathbf{x}^*)$, defined in a surface α and such that

$$\delta V^* = \mathbf{p}^* . \delta\mathbf{x}^*, \tag{2.6.11}$$

where $\mathbf{p}^*$ denotes the ray vector of a member of a normal congruence where it intersects α, $V(\mathbf{x})$ may be found as follows. If $V^\dagger(\mathbf{x}^*, \mathbf{x})$ is the point characteristic function,

$$\delta V^\dagger = \mathbf{p}.\delta\mathbf{x} - \mathbf{p}^*.\delta\mathbf{x}^*. \tag{2.6.12}$$

If we now identify the arguments of the characteristic function with the co-ordinates of the points P^*, a point of α, and P which lies on the ray of the normal congruence which passes through P^*,

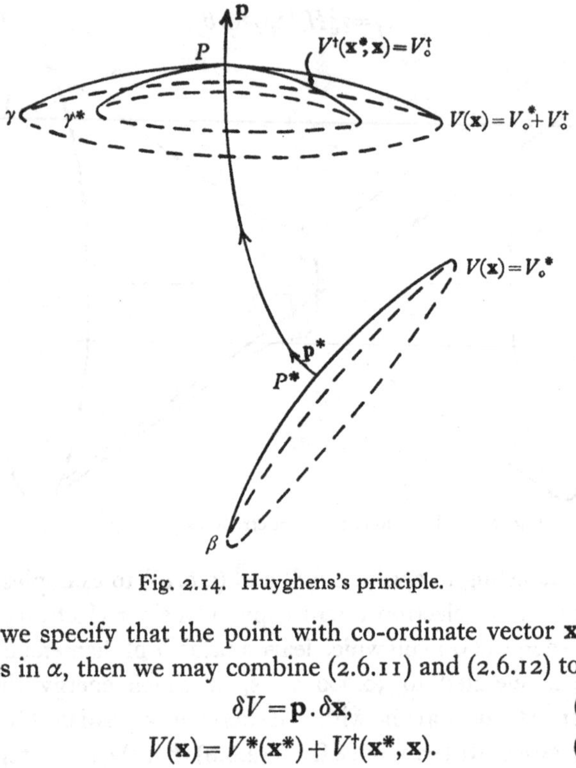

Fig. 2.14. Huyghens's principle.

and if we specify that the point with co-ordinate vector $\mathbf{x}^* + \delta\mathbf{x}^*$ also lies in α, then we may combine (2.6.11) and (2.6.12) to obtain

$$\delta V = \mathbf{p}.\delta\mathbf{x}, \tag{2.6.13}$$

where $$V(\mathbf{x}) = V^*(\mathbf{x}^*) + V^\dagger(\mathbf{x}^*, \mathbf{x}). \tag{2.6.14}$$

Hence (2.6.14) gives the function $V(\mathbf{x})$. In this formula, $V^\dagger(\mathbf{x}^*, \mathbf{x})$ may be evaluated by a Hilbert integral, since its terminal points are connected by rays of the normal congruence.

Huyghens's principle. If we assume that $V^* = V_0^*$, a constant, the ray vectors of the congruence are orthogonal to α. Moreover, the surface defined by $V^\dagger(\mathbf{x}^*, \mathbf{x}) = V_0^\dagger$, where $\mathbf{x}^*$ is fixed, is orthogonal to the ray leaving P^* and passes through P, where it is orthogonal to the vector $\mathbf{p}$ (fig. 2.14).

Let us now suppose that P^* is varied so that $\mathbf{x}^*$ is changed to $\mathbf{x}^*+\delta\mathbf{x}^*$. If P^* is constrained to remain in the surface α, $\mathbf{p}^*.\delta\mathbf{x}^*=0$ so that (2.6.12) gives $\mathbf{p}.\delta\mathbf{x}=0$ and the displaced orthogonal surface still passes through P and is normal to $\mathbf{p}$ at this point.

The result may be stated as follows: *If α is a surface orthogonal to a given normal congruence, if through each point P^* of α there passes a subsidiary normal congruence, and if we consider the surfaces β^* which are orthogonal to these congruences, the optical distance—measured along a ray—of each such surface to its base point being the same, then the envelope β of the surfaces β^* is an orthogonal surface of the original normal congruence. Moreover, the ray of this congruence which passes through a point P^* of α intersects β at P, the point of contact of β^* and β.*

The above statement is recognized to be a geometrical formulation of Huyghens's principle.‡ The surfaces orthogonal to normal congruences are often referred to, in geometrical optics, as 'wave surfaces', since *one may pass from geometrical optics to wave optics by interpreting the wave surfaces as surfaces of constant phase.* §

2.7. The Hamilton-Jacobi equation

It was seen in the last section that a normal congruence may be described by a modified form of the point characteristic function, which we have named the 'optical potential', which is a function of only one position vector. However, we should expect that the physical laws of geometrical optics will impose some restriction upon this function or—taking another viewpoint—that there is an equation involving this function which is equivalent to our original variational equation. In this section we shall obtain this new formulation of the basic principle of electron optics and investigate some of its consequences.

Let us consider the function $\mathscr{H}(\mathbf{x},\mathbf{p})$ defined by

$$\mathscr{H}=\tfrac{1}{2}\{p^{-1}(\mathbf{p}+\mathbf{A})^2-p\}, \tag{2.7.1}$$

where, in the present context, the position and ray vectors are regarded as independent.‖ We find, on differentiating and using (2.2.8), that

$$\frac{\partial\mathscr{H}}{\partial\mathbf{p}}=\mathbf{l}. \tag{2.7.2}$$

‡ Cf. C. Caratheodory, *Geometrische Optik* (Springer, Berlin, 1937), p. 13.

§ For further details see, for instance, Hopkins, op. cit.

‖ $\mathbf{p}$ and $\mathbf{x}$ are normally related by $(\mathbf{p}+\mathbf{A}(\mathbf{x}))^2=p(\mathbf{x})^2$, which follows from (2.2.8).

On differentiating by $\mathbf{x}$ we find that

$$\frac{\partial \mathscr{H}}{\partial x_r} = -\frac{\partial p}{\partial x_r} + l_s \frac{\partial A_s}{\partial x_r}, \tag{2.7.3}$$

the right-hand side of which is clearly equal to $-\partial n/\partial x_r$ and therefore, by virtue of the ray equation (2.2.10), to $-\mathrm{d}p_r/\mathrm{d}s$. Equations (2.7.2) and (2.7.3) may therefore be written as

$$\frac{\mathrm{d}\mathbf{x}}{\mathrm{d}s} = \frac{\partial \mathscr{H}}{\partial \mathbf{p}}, \quad \frac{\mathrm{d}\mathbf{p}}{\mathrm{d}s} = -\frac{\partial \mathscr{H}}{\partial \mathbf{x}}. \tag{2.7.4}$$

We now recognize (2.7.1) to be the *Hamiltonian function* and (2.7.4) the *canonical equations* of our system. It should be noted that, since we are now regarding both $\mathbf{x}$ and $\mathbf{p}$ as independent variables, the first of equations (2.7.4) is valid even in field-free space. Moreover, it is an immediate consequence of (2.7.4) that $\mathrm{d}\mathscr{H}/\mathrm{d}s$ vanishes so that $\mathscr{H}$ has a constant value which, from (2.2.8), is obviously zero.

If we now use (2.6.3) to substitute $\partial V/\partial \mathbf{x}$ for $\mathbf{p}$ in (2.7.1) and equate the Hamiltonian to zero, we obtain

$$\left\{\frac{\partial V}{\partial \mathbf{x}} + \mathbf{A}\right\}^2 = p^2, \tag{2.7.5}$$

which is the *Hamilton-Jacobi* equation‡ for our system. It is not difficult to demonstrate that this is the equation we seek by showing that it leads back to both the variational equation and the ray equation.

Let us now suppose that $V^*(\mathbf{x})$ is a solution of (2.7.5), that $\mathbf{p}^*(\mathbf{x})$ is the ray-vector distribution then defined by (2.6.3) and that $\mathbf{l}^*(\mathbf{x})$ is the direction-vector distribution then defined by (2.2.8). On noting that $\mathbf{A} = p\mathbf{l}^* - \partial V^*/\partial \mathbf{x}$, we may deduce from (1.3.9) and (1.3.10) that

$$\delta \int_A^B n\,\mathrm{d}s = \delta \int_A^B \left\{p - \mathbf{l}.\left(p\mathbf{l}^* - \frac{\partial V^*}{\partial \mathbf{x}}\right)\right\} \mathrm{d}s. \tag{2.7.6}$$

The last term of the integrand may be integrated exactly and its variation vanishes since the end-points are fixed. Hence

$$\delta \int_A^B n\,\mathrm{d}s = \int_A^B \left\{(1 - \mathbf{l}.\mathbf{l}^*)\left(\delta p + p\frac{\mathrm{d}\,\delta s}{\mathrm{d}s}\right) - p(\mathbf{l}.\delta\mathbf{l}^* + \mathbf{l}^*.\delta\mathbf{l})\right\} \mathrm{d}s, \tag{2.7.7}$$

‡ See, for instance, ref. (12), pp. 233 et seq.

from which it is clear that the variation of the integral on the left-hand side vanishes identically if $\mathbf{l}(\mathbf{x})=\mathbf{l}^*(\mathbf{x})$. This shows that the Hamilton-Jacobi equation is equivalent to the variational equation with which we began.

Spatial differentiation of (2.7.5) gives

$$\left\{\frac{\partial V}{\partial x_r}+A_r\right\}\left\{\frac{\partial^2 V}{\partial x_r\,\partial x_s}+\frac{\partial A_r}{\partial x_s}\right\}=p\frac{\partial p}{\partial x_s}, \tag{2.7.8}$$

which, with the help of (2.2.8) and (2.6.3), may be rewritten as

$$l_r\frac{\partial p_s}{\partial x_r}=\frac{\partial p}{\partial x_s}-l_r\frac{\partial A_r}{\partial x_s}. \tag{2.7.9}$$

We now see that, since (2.7.9) is identical with (2.2.10), it is possible to derive the ray equation from the Hamilton-Jacobi equation.

We have so far restricted our attention to one particular solution of the equation (2.7.5), but let us now consider a family of solutions represented by the function $V(\mathbf{x},\alpha_i)$ which involves two parameters α_1, α_2 in such a way that

$$\begin{vmatrix} \dfrac{\partial^2 V}{\partial\alpha_1\,\partial x_r} & \dfrac{\partial^2 V}{\partial\alpha_1\,\partial x_s} \\ \dfrac{\partial^2 V}{\partial\alpha_2\,\partial x_r} & \dfrac{\partial^2 V}{\partial\alpha_2\,\partial x_s} \end{vmatrix} \neq 0 \tag{2.7.10}$$

for some values of r and s for all $\mathbf{x}$, i.e. that $\partial^2 V/\partial\alpha_1\,\partial\mathbf{x}$ and $\partial^2 V/\partial\alpha_2\,\partial\mathbf{x}$ are not parallel vectors. It is seen that the condition (2.7.10) precludes one of the parameters being an additive constant.

On differentiating (2.7.5) by α_i, we obtain

$$\left\{\frac{\partial V}{\partial x_r}+A_r\right\}\frac{\partial^2 V}{\partial x_r\,\partial\alpha_i}=0, \tag{2.7.11}$$

from which, on referring to (2.2.8) and (2.6.3), we see that

$$\frac{\mathrm{d}}{\mathrm{d}s}\frac{\partial V}{\partial\alpha_i}=0. \tag{2.7.12}$$

It follows that, along any ray of the system,

$$\frac{\partial V}{\partial\alpha_i}=\beta_i, \tag{2.7.13}$$

where β_1, β_2 are constant for that ray.

Since it is a consequence of the condition (2.7.10) that the pair of surfaces defined by (2.7.13) cannot possess a common normal at any point of intersection of the surfaces, the surfaces must intersect in a curve which must be a ray. This ray, being determined by the four parameters $\alpha_1, \alpha_2, \beta_1, \beta_2$, is the typical member of a quadruple manifold which we may identify with the complete set of rays of the system. Hence *if $V(\mathbf{x}, \alpha_i)$ is any complete solution of the Hamilton-Jacobi equation* (2.7.5), *the totality of rays may be found by solving the equation* (2.7.13). The solution would be required in the form $\mathbf{x}=\mathbf{x}(\alpha_i, \beta_i, s)$. This result is often referred to as the *Jacobi-Liouville theorem.*

The point characteristic function $V(\mathbf{x}_a, \mathbf{x}_b)$, which was introduced in §2.2, may be regarded as a special form of the solution $V(\mathbf{x}, \alpha_i)$ of the Hamilton-Jacobi equation, for we may interpret $\mathbf{x}_b$ as $\mathbf{x}$ and x_{ia} as α_i, putting $x_{3a}=0$. We shall now prove that, conversely, it is possible to derive the point characteristic function $V(\mathbf{x}_a, \mathbf{x}_b)$ from a complete solution of the Hamilton-Jacobi equation.

If the complete solution of (2.7.5) is written as $V^*(\mathbf{x}, \alpha_i)$, we see that, since (2.7.13) holds along the ray joining A and B,

$$\frac{\partial V^*}{\partial \alpha_i}=\beta_i \quad \text{for} \quad \mathbf{x}=\mathbf{x}_a \quad \text{and for} \quad \mathbf{x}=\mathbf{x}_b. \qquad (2.7.14)$$

We may suppose that the four equations (2.7.14) are solved for the four parameters α_i, β_i which are thereby expressed as functions of $\mathbf{x}_a$, $\mathbf{x}_b$. The function defined by

$$V=V^*(\mathbf{x}_b, \alpha_i)-V^*(\mathbf{x}_a, \alpha_i) \qquad (2.7.15)$$

may now be expressed in the form $V(\mathbf{x}_a, \mathbf{x}_b)$; it is obvious that $V(\mathbf{x}_a, \mathbf{x}_a)=0$. On differentiating (2.7.15), we find that

$$\frac{\partial V}{\partial \mathbf{x}_b}=\frac{\partial V^*}{\partial \mathbf{x}}(\mathbf{x}_b, \alpha_j)+\frac{\partial \alpha_i}{\partial \mathbf{x}_b}\frac{\partial V^*}{\partial \alpha_i}(\mathbf{x}_b, \alpha_j)-\frac{\partial \alpha_i}{\partial \mathbf{x}_b}\frac{\partial V^*}{\partial \alpha_i}(\mathbf{x}_a, \alpha_j). \qquad (2.7.16)$$

The second and third terms on the right-hand side cancel by virtue of (2.7.14); moreover, since α_i was found from (2.7.14), the first term represents the ray vector $\mathbf{p}_b$ for the ray connecting A to B. This shows that V satisfies the second of equations (2.2.16); that it satisfies the first equation may be shown in the same manner. It is easy to verify that the Hamilton-Jacobi equation may be transformed into Hamilton's partial differential equations (2.2.18) by application of (2.7.16) and its counterpart for the point A. We now see that *it is possible to solve Hamilton's pair of partial differential equations by finding any complete solution of the Hamilton-Jacobi equation.*

It should be remarked, in parentheses, that although it is possible to solve a system completely by finding a complete solution of the Hamilton-Jacobi equation, one is often content to find only a particular normal congruence traversing the system. Thus if we were examining a β-ray spectrometer operating at wide aperture, we could very conveniently trace the normal congruence which passes through a point object by solving the equation (2.7.5). We shall see in later paragraphs that it is sometimes possible to take advantage of the symmetry of a system in such a way as to obtain a set of neighbouring congruences from only one solution of the Hamilton-Jacobi equation.

Let us consider the form taken by (2.7.5) in an instrument of rotational symmetry. If we adopt cylindrical co-ordinates (z, r, θ), the magnetic vector potential may be taken to have only one non-zero component, A_θ (cf. § 1.5). Equation (2.7.5) now becomes

$$\left\{\frac{\partial V}{\partial z}\right\}^2 + \left\{\frac{\partial V}{\partial r}\right\}^2 + \left\{\frac{1}{r}\frac{\partial V}{\partial \theta} + A_\theta\right\}^2 = p^2, \qquad (2.7.17)$$

wherein $p = p(z, r)$ and $A_\theta = A_\theta(z, r)$. It is easy to see that a one-parameter family of solutions of (2.7.17) is given by

$$V(z, r, \theta; C) = U(z, r; C) + C\theta, \qquad (2.7.18)$$

where C is the parameter and U satisfies the equation

$$\left\{\frac{\partial U}{\partial z}\right\}^2 + \left\{\frac{\partial U}{\partial r}\right\}^2 = p^2 - (A_\theta + C/r)^2. \qquad (2.7.19)$$

It is clear that our three-dimensional problem has now been reduced to one of two dimensions. The 'z-r' or *meridional* representation is characterized by a *meridional refractive index* $\sqrt{(p^2 - (A_\theta + C/r)^2)}$, which is isotropic.

The significance of the parameter C becomes apparent when we obtain the ray vector associated with the function (2.7.18). We see from (2.6.3) that

$$p_z = \frac{\partial U}{\partial z}, \quad p_r = \frac{\partial U}{\partial r}, \quad p_\theta = \frac{C}{r}, \qquad (2.7.20)$$

and hence from (2.2.8) that

$$pl_z = \frac{\partial U}{\partial z}, \quad pl_r = \frac{\partial U}{\partial r}, \quad pl_\theta = A_\theta + \frac{C}{r}. \qquad (2.7.21)$$

The result may be stated as follows: *The quantity* rp_θ, *i.e.* $r(pl_\theta - A_\theta)$, *is constant along any ray in an instrument of rotational symmetry; moreover, the triple manifold of rays for which this constant takes a particular value C may be examined in the meridional plane as a double manifold;*‡ *the meridional refractive index is isotropic and is given by the formula* $\sqrt{(p^2-(A_\theta+C/r)^2)}$.

The reader may care to turn back to § 1.5 and verify that (1.5.4) is a statement of the relation $rp_\theta = C$ and that (1.5.7) is the ray equation derived from the meridional refractive index when the field is purely magnetic.

If we now refer to equation (2.7.13), we find that, for any ray, U satisfies the further partial differential equation

$$\frac{\partial U}{\partial C} = \theta_0 - \theta, \tag{2.7.22}$$

where θ_0 is a constant. Since $\theta(z, r; C)$ may be derived from (2.7.21) —the result being the integrated form of (1.5.5)—we may apply (2.7.22) to calculate the effect of a small change of the parameter C, for

$$\delta U(z, r; C) = \delta C(\theta_0 - \theta(z, r; C)). \tag{2.7.23}$$

This equation might possibly prove useful if, having traced an assembly of non-skew rays (for which $C=0$) through a bounded field of rotational symmetry, one wished to study an assembly of slightly skew rays (for which C is non-zero but small). If one specifies the relation between the two assemblies on the 'object' side of the field, equation (2.7.23) and the perturbed form of (2.7.21) determine θ_0 and δC for each ray. Since the perturbed (skew) rays, to which (2.7.23) strictly applies, differ only slightly from the unperturbed (non-skew) rays, we may apply (2.7.23) to the latter and so determine δU on the 'image' side of the field. The direction of the skew rays, where they emerge from the field, is now given by the perturbed form of (2.7.21). This procedure appears to be equivalent to the use of the appropriate perturbation characteristic function, but rather more complicated.

Let us now make a brief examination of the perturbation of electron-optical systems in the light of the Hamilton-Jacobi

‡ It is because the triple manifold is rotationally symmetrical that it is reduced to a double manifold on projection into the meridional plane.

equation. The notation of § 2.4 will be used except that (2.4.1) will be extended to the second order:

$$\mathsf{P}n = n + \varpi n^{\mathrm{I}} + \varpi^2 n^{\mathrm{II}} + \dots. \tag{2.7.24}$$

We shall also need the perturbations of (2.2.8) and (2.6.3), which are found to be

$$\left.\begin{aligned} \mathbf{p}^{\mathrm{I}} &= p\mathbf{l}^{\mathrm{I}} + p^{\mathrm{I}}\mathbf{l} - \mathbf{A}^{\mathrm{I}}, \\ \mathbf{p}^{\mathrm{II}} &= p\mathbf{l}^{\mathrm{II}} + p^{\mathrm{I}}\mathbf{l}^{\mathrm{I}} + p^{\mathrm{II}}\mathbf{l} - \mathbf{A}^{\mathrm{II}}, \quad \text{etc.} \end{aligned}\right\} \tag{2.7.25}$$

and

$$\mathbf{p}^{\mathrm{I}} = \frac{\partial V^{\mathrm{I}}}{\partial \mathbf{x}}, \quad \mathbf{p}^{\mathrm{II}} = \frac{\partial V^{\mathrm{II}}}{\partial \mathbf{x}}, \quad \text{etc.} \tag{2.7.26}$$

The Hamilton-Jacobi equation (2.7.5) may be obtained by eliminating $\mathbf{l}$ from the identity $\mathbf{l}^2 = 1$ by means of (2.2.8) and then eliminating $\mathbf{p}$ by means of (2.6.3). We shall deal with perturbations by an extension of this recipe, starting from

$$\mathbf{l}.\mathbf{l}^{\mathrm{I}} = 0, \quad (\mathbf{l}^{\mathrm{I}})^2 + 2\mathbf{l}.\mathbf{l}^{\mathrm{II}} = 0, \quad \text{etc.} \tag{2.7.27}$$

The first equations of the above three sets give

$$\mathbf{l}.\left(\frac{\partial V^{\mathrm{I}}}{\partial \mathbf{x}} - p^{\mathrm{I}}\mathbf{l} + \mathbf{A}^{\mathrm{I}}\right) = 0, \tag{2.7.28}$$

which may be integrated at once to give

$$V^{\mathrm{I}}(\mathbf{x}_a, \mathbf{x}_b) = \int_A^B (p^{\mathrm{I}} - \mathbf{l}.\mathbf{A}^{\mathrm{I}})\, ds, \tag{2.7.29}$$

where the integral is evaluated along the unperturbed ray connecting A and B. The formulation (2.7.29), in which the constant of integration has been eliminated, identifies the result with that obtained in § 2.4.

However, there is no difficulty in proceeding by the present method to the second and higher orders. The second equations of (2.7.27), (2.7.28) and (2.7.29) give

$$p(\mathbf{l}^{\mathrm{I}})^2 + 2\mathbf{l}.\left(\frac{\partial V^{\mathrm{II}}}{\partial \mathbf{x}} - p^{\mathrm{II}}\mathbf{l} + \mathbf{A}^{\mathrm{II}}\right) = 0, \tag{2.7.30}$$

which, upon integration, yields the formula

$$V^{\mathrm{II}}(\mathbf{x}_a, \mathbf{x}_b) = \int_A^B (p^{\mathrm{II}} - \mathbf{l}.\mathbf{A}^{\mathrm{II}} - \tfrac{1}{2}p(\mathbf{l}^{\mathrm{I}})^2)\, ds, \tag{2.7.31}$$

the integral again being evaluated along the unperturbed ray.

If the perturbation involves two parameters, so that (2.7.24) should be replaced by

$$\mathsf{P}n = n + \varpi n^{\mathrm{I}} + \rho n^{J} + \varpi^2 n^{\mathrm{II}} + \varpi\rho n^{\mathrm{I}J} + \rho^2 n^{JJ} + \dots, \tag{2.7.32}$$

the 'mixed' perturbation characteristic function of the second order is found to be given by

$$V^{IJ}(\mathbf{x}_a, \mathbf{x}_b) = \int_A^B (p^{IJ} - \mathbf{l}.\mathbf{A}^{IJ} - p\mathbf{l}^I.\mathbf{l}^J)\, ds. \qquad (2.7.33)$$

The formulae (2.7.31) and (2.7.33) could not be obtained with such ease by the method of §2.4, but, on the other hand, the present method gives only the *point* perturbation characteristic functions, throwing no light on the general form of the differential perturbation relations (see (2.4.10)).

The above formulae throw interesting light upon the variational equation. If the field suffers a perturbation depending on the parameter ϖ, the refractive index may be written as $n(\varpi)$; similarly, the ray connecting A and B may be denoted by $\Gamma(\varpi)$. We now see that

$$\int_{\Gamma(\varpi)} n(\varpi)\, ds = V + \varpi V^I + \varpi^2 V^{II} + O(\varpi^3). \qquad (2.7.34)$$

If we now compare the integral in the perturbed field along the unperturbed ray with the integral in the perturbed field along the perturbed ray, using (2.7.29) and (2.7.31), we see that

$$\int_{\Gamma(\varpi)} n(\varpi)\, ds = \int_{\Gamma(0)} n(\varpi)\, ds - \tfrac{1}{2}\varpi^2 \int_{\Gamma(0)} p(\mathbf{l}^I)^2\, ds + O(\varpi^3). \qquad (2.7.35)$$

If—as is permissible—we now interchange our designation of the 'perturbed' and 'unperturbed' rays and work only to the second order in ϖ, we see from (2.7.35) that

$$\int_{\Gamma(0)} n(0)\, ds < \int_{\Gamma(\varpi)} n(0)\, ds, \qquad (2.7.36)$$

a result which may be stated as follows: *If we compare the optical length of a ray with the optical lengths of neighbouring paths, with the same terminal points, into which the ray may be displaced by appropriate perturbations of the field, the optical length of the ray is a true minimum.*

This result may be contrasted with the usual formulation of the variational equation—which embraces Fermat's principle for isotropic media—for which the optical length along the ray takes a stationary value but is not necessarily a minimum.‡

‡ See, for example, ref. (11), p. 96.

CHAPTER 3

INSTRUMENTAL ELECTRON OPTICS

3.1. Introduction

We saw in Chapter 1 that the laws of geometrical electron optics may be formulated as a variational principle (1.3.4). Although the equation was expressed in terms of rectangular Cartesian co-ordinates, it is an advantage of the variational formulation that the equation may be applied with facility to problems which require the use of curvilinear co-ordinates. It is sometimes convenient, as in §1.4, to begin with rectangular co-ordinates and then transform to curvilinear co-ordinates (cf. (1.4.27)); on the other hand, we may use curvilinear co-ordinates from the start, as in §1.5, although transformations (such as (1.5.10)) may follow.

In Chapter 2, which was devoted to a presentation of the concepts of classical optics, it was expedient to restrict ourselves, with minor exceptions, to rectangular co-ordinates. However, in examining the properties of a specific instrument it may, as we have already seen, be advisable to work with curvilinear co-ordinates. The title of this chapter is intended to imply that our present aim is to set out the principal methods which may be applied to the analysis of the properties of an electron-optical instrument; we should therefore begin by introducing an appropriate general set of curvilinear co-ordinates and referring the variational equation to this set.

It was seen in §§1.4 and 1.5 that it is possible to obtain perfect imaging in a rotationally symmetrical system only to the paraxial approximation and for a monochromatic beam. It will be shown in §3.3 that it is also possible for a beam which follows a *curved* path to form a good image provided that the beam is sufficiently narrow. By introducing curvilinear co-ordinates based upon a *ray axis* which follows a ray of the beam, the other two co-ordinates measuring displacements from this axis, it is again possible to define a 'paraxial approximation' according to which an instrument may exhibit perfect imaging properties.

Since the paraxial properties of electron-optical systems usually predominate over higher-order effects, the optical properties which beams exhibit to the paraxial approximation are of great interest and they are treated in § 3.4 under their customary title of 'Gaussian dioptrics'. Complications arise in the electron-optical theory which are not found in light optics, since, in the former, the object and image are not necessarily in field-free space.

On looking back to (1.4.14), we see that it may be possible, when considering the paraxial behaviour of rays, to treat the off-axis co-ordinates separately. It is shown in § 3.3 that it is possible to put the paraxial variational function into 'separated' or *orthogonal* form only if a certain condition is satisfied. If the paraxial system is orthogonal, the off-axis co-ordinates of a ray will generally satisfy different ray equations, but it is possible to apply Gaussian dioptrics by considering one co-ordinate at a time.

As has already been stated in § 1.5, it is not enough to calculate the paraxial image-forming properties of an electron-optical instrument; one must also calculate the accuracy with which these properties will in practice be realized. This, as we saw in § 2.1, is a perturbation problem, so that we should reintroduce the method of perturbation characteristic functions, expressing it in terms of the formalism of the present chapter. Since most instruments contain an aperture or 'stop', it is convenient to specify rays by their co-ordinates in the object and aperture planes. It follows that it would be desirable to use the same co-ordinates as the arguments of the characteristic functions. An interesting consequence of this choice is that the first-order perturbation of an assembly of rays is characterized by a *pair* of characteristic functions, although one of the characteristic functions drops out if the perturbation is required only in the *image* plane of the unperturbed system, i.e. in the paraxial or 'Gaussian' image plane.

3.2. Curvilinear co-ordinates

In analysing an electron-optical instrument, it is necessary to proceed along lines similar to those adopted in §§ 1.4 and 1.5 so as to bring to light the dependence of the imaging properties upon the narrowness of the beam. The paraxial or 'first-order' approximation which, as we have seen, may give perfect imaging, is approached as the beam width is reduced indefinitely. It will therefore be necessary, in examining an instrument which employs a *curved* beam, to adopt co-ordinates whose magnitudes are directly related to the width of the electron beam.

It is proposed that we retain the notation (x_i, z), but from now on, unless the contrary is specifically stated, regard these as a *curvilinear* set. The z-axis, i.e. the curve $x_i = 0$, will be taken to be a ray of the beam and referred to as the *ray axis*. The 'off-axis' co-ordinates x_1, x_2 may complete the determination of points in the neighbourhood of the axis in any manner, but we shall confine our attention to the choice depicted in fig. 3.1; for each value of z, the co-ordinates (x_1, x_2) form a plane rectangular Cartesian set, the plane being normal to the ray axis.

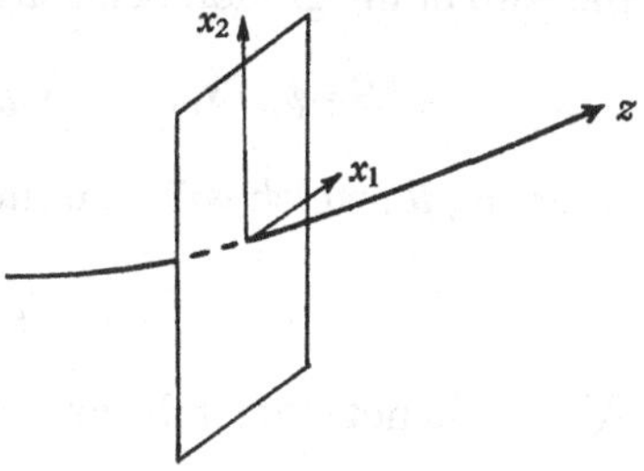

Fig. 3.1. Curvilinear co-ordinates based on a 'ray axis'.

In the present section we shall indicate briefly how the theory which was developed in the last chapter may be expressed in terms of curvilinear co-ordinates. We need therefore make no assumptions about the beam width for the time being.

In terms of our new co-ordinates, the variational equation (1.3.9) may be expressed as

$$\delta \int_A^B m \, \mathrm{d}z = 0, \tag{3.2.1}$$

where

$$m = n \frac{\mathrm{d}s}{\mathrm{d}z}. \tag{3.2.2}$$

This function may be expressed as $m(x_i, x_i', z)$, where a prime henceforth denotes differentiation with respect to z; so expressed, it will be known as the *variational function*.

We saw in § 1.3 that it is possible to add to the vector potential the gradient of a scalar without affecting the significance of the equation (1.3.4). Similarly, the interpretation of *a variational function* by means of (3.2.1) *is unaltered if a total differential is added to the function*, i.e. if m is changed to $m + \mathrm{d}f/\mathrm{d}z$, where $f = f(x_i, z)$. The proof is trivial. A useful alternative statement is the following: *If the variational function may be expressed as a sum of functions, one of which is of the form* $\mathrm{d}f/\mathrm{d}z \,.\, g$, *where* f *and* g *are functions of* x_i *and* z *only, then this term may be replaced by* $-f \,.\, \mathrm{d}g/\mathrm{d}z$. This will be quoted as the *partial integration rule*.

Let us now draw a parallel to § 2.2 in the present notation but confine our attention to variations which may be written as $\delta x_i(z)$, so precluding possible displacements of the reference surfaces, a restriction which is of no importance. If

$$S = \int_A^B m\,\mathrm{d}z, \tag{3.2.3}$$

the path of integration being arbitrary, we find that

$$\delta S = p_{ib}\,\delta x_{ib} - p_{ia}\,\delta x_{ia} - \int_{z_a}^{z_b} \delta x_i \left\{\frac{\mathrm{d}p_i}{\mathrm{d}z} - \frac{\partial m}{\partial x_i}\right\} \mathrm{d}z, \tag{3.2.4}$$

where p_1, p_2, which will be termed the *ray variables*, are defined by

$$p_i = \frac{\partial m}{\partial x_i'}. \tag{3.2.5}$$

We should note that if the total differential $\dfrac{\mathrm{d}}{\mathrm{d}z} f(x_i, z)$ is added to m, the term $\partial f/\partial x_i$ should be added to p_i.

Since, according to (3.2.1), δS vanishes for all $\delta x_i(z)$ if the path of integration is a ray and if $\delta x_{ia} = \delta x_{ib} = 0$, we obtain from (3.2.4) the *ray equation*

$$\frac{\mathrm{d}p_i}{\mathrm{d}z} = \frac{\partial m}{\partial x_i}, \tag{3.2.6}$$

and the new form of *Hamilton's differential relation*

$$\delta V_{ab} = p_{ib}\,\delta x_{ib} - p_{ia}\,\delta x_{ia}, \tag{3.2.7}$$

where it is convenient to write $V_{ab}(x_{ia}, x_{ib})$ in place of $V(x_{ia}, z_a, x_{ib}, z_b)$ for Hamilton's *point characteristic function* defined by

$$V_{ab} = \int_{z_a}^{z_b} m\,\mathrm{d}z, \tag{3.2.8}$$

the integral now being evaluated along a ray. It is seen that (3.2.6) is, by virtue of (3.2.5), identical with (1.4.7) which was quoted as the Euler-Lagrange equation derivable from (3.2.1).

If the increments δx_{ia} and δx_{ib} are unrelated, i.e. in the absence of stigmatic imaging between the planes $z = z_a$ and $z = z_b$, we may deduce from (3.2.7) that

$$p_{ia} = -\frac{\partial V_{ab}}{\partial x_{ia}} \quad \text{and} \quad p_{ib} = \frac{\partial V_{ab}}{\partial x_{ib}}, \tag{3.2.9}$$

which are of the same form as (2.2.16). Now V_{ab}, defined by (3.2.8), and V, defined by (2.2.11), are equal although they may be referred to different sets of co-ordinates. However, if we may take x_{ia} and x_{ib} to refer to the same co-ordinates in both (2.2.16) and (3.2.9), p_{ia} and p_{ib} must have the same values in each case. Hence, *if the pair of co-ordinates* x_1, x_2 *form a Cartesian set, the ray variables* p_1, p_2 *defined by* (3.2.5) *are identical with the corresponding components of the ray vector defined by* (2.2.8). One may note that the addition of the term $\mathrm{d}f/\mathrm{d}z$ to m is equivalent to the addition of grad χ to $\mathbf{A}$, where $\chi = -f$.

There would be no difficulty in redefining such concepts as the 'Lagrange invariant', 'normal congruences' and the 'Hamiltonian' in terms of curvilinear co-ordinates, but there is little incentive to pursue this course. Examples of the use of the Hamiltonian in electron-optical calculations may be found in the publications of Chaco and Blank‡ and Cotte.§

3.3. The paraxial approximation

We saw in §§ 1.4 and 1.5 that instruments of rotational symmetry display perfect imaging properties provided that the beam is monochromatic and narrow enough for the 'paraxial approximation' to be effectively realized. We shall investigate the optical properties associated with such image formation in the next section, but we shall first proceed to demonstrate, in the present section, that systems with curved optical axes may display the same paraxial imaging properties as systems of rotational symmetry.

We adopt the curvilinear co-ordinates shown in fig. 3.1 and expand the variational function in the form

$$m = m^{(0)} + m^{(1)} + m^{(2)} + m^{(3)} + \dots, \tag{3.3.1}$$

where $m^{(r)}$ denotes a homogeneous polynomial of the rth degree in x_i and x_i', the coefficients being functions of z. Obviously the term $m^{(0)}$ may be ignored. By the partial integration rule, the term involving x_i' in $m^{(1)}$ may be absorbed into the term involving x_i. However, it is easily seen from (3.2.5) and (3.2.6) that the condition that the z-axis should be a ray is that the remaining coefficient in $m^{(1)}$ should vanish so that $m^{(1)}$ also may be cancelled.

‡ Ref. (11): Supplementary Note I by N. Chaco and A. A. Blank.
§ Ref. (36), pp. 373 et seq.

It is now seen that as the arguments x_i and x_i' become indefinitely small m tends to $m^{(2)}$. In the neighbourhood of a *cusp*, i.e. of a point of reflexion, the derivative x_i' does not tend to zero, for all z, as x_i tends to zero. The treatment of electron mirrors must therefore be distinguished from the treatment of purely refractive fields, and it is necessary to exclude the former from subsequent considerations. *The paraxial approximation*, the consequences of which we shall investigate in this section, *is the approximation involved in replacing the variational function by its second-order term.*

It would seem at first sight that $m^{(2)}$ must be a linear combination of the terms $x_1'^2$, $x_1'x_2'$, $x_2'^2$, $x_1'x_1$, $x_1'x_2$, $x_2'x_1$, $x_2'x_2$, x_1^2, x_1x_2 and x_2^2. However, the terms $x_1'x_1$ and $x_2'x_2$ may be absorbed into x_1^2 and x_2^2 by means of the partial integration rule and, if the terms $x_1'x_2$ and $x_2'x_1$ are rearranged as a combination of $(x_1'x_2 - x_2'x_1)$ and $(x_1'x_2 + x_2'x_1)$, the latter may be absorbed into the term x_1x_2. Let us now consider the first three terms of the set; it is not difficult to see from (1.3.10) and (3.2.2) that these all originate in the term $p\,ds/dz$. We also see that if $x_i = 0$, $ds^2 = dz^2 + dx_i^2$; and if $dz = 0$, $ds^2 = dx_i^2$. It follows that the metric is of the form

$$ds^2 = (1 + O(x_i))\,dz^2 + dx_i^2 + O(x_i)\,dz\,dx_i, \qquad (3.3.2)$$

so that the terms of interest appear in $m^{(2)}$ as the combination $\frac{1}{2}\mathsf{p}x_i'^2$, where $\mathsf{p}(z) = p(0, z)$.

The general form of $m^{(2)}$ may now be written as

$$m^{(2)} = \tfrac{1}{2}\mathsf{p}(x_1'^2 + x_2'^2) + \mathsf{t}(x_1'x_2 - x_1x_2') + \tfrac{1}{2}\mathsf{e}_1x_1^2 + \mathsf{f}x_1x_2 + \tfrac{1}{2}\mathsf{e}_2x_2^2, \qquad (3.3.3)$$

where t, e_1, e_2 and f also are functions of z. It is worth noting that there are two contributions to the term t: one from the metric by way of the third term on the right-hand side of (3.3.2), and another from the magnetic field by way of the term $-\mathbf{1}.\mathbf{A}\,ds/dz$. In choosing the transformation (1.5.10), we were arranging for the former to balance the latter.

There is no difficulty in writing down the formulae for the ray variables and the ray equations derivable from (3.3.3) by means of (3.2.5) and (3.2.6), but it will serve our purpose merely to note that each formula or equation involves both off-axis co-ordinates. It follows that we cannot expect the system described by (3.3.3) to display imaging properties similar to those obtained in §§ 1.4 and

1.5, for in the ray equations (1.4.12) and (1.5.13) x_1 and x_2 and the real and imaginary parts of v may be considered independently. Our aim, therefore, is to discover whether by a transformation of co-ordinates the coefficients t and f may be eliminated from (3.3.3).

Let us suppose that by a transformation from co-ordinates x_i to co-ordinates y_i we may put $m^{(2)}$ into the separated or *orthogonal* form

$$m^{(2)} = \tfrac{1}{2}(\mathsf{p}y_1'^2 - \mathsf{Y}_1 y_1^2) + \tfrac{1}{2}(\mathsf{p}y_2'^2 - \mathsf{Y}_2 y_2^2). \tag{3.3.4}$$

The paraxial ray variables are then given by

$$p_{yi} = \mathsf{p}y_i', \tag{3.3.5}$$

and the paraxial ray equations are

$$\frac{d}{dz}\left(\mathsf{p}\frac{dy_1}{dz}\right) + \mathsf{Y}_1 y_1 = 0, \quad \frac{d}{dz}\left(\mathsf{p}\frac{dy_2}{dz}\right) + \mathsf{Y}_2 y_2 = 0. \tag{3.3.6}$$

We shall now consider what condition must be satisfied in order that (3.3.3) may be put into the form (3.3.4).

Fortunately there are restrictions on the transformations it is profitable to consider. If the separation can be effected at all, it can be done by a transformation which does not change the scale as is obvious, for, once the separation is achieved, magnification of the co-ordinates cannot create 'mixed' terms but it can restore the original scale. It is also seen that we should neither adopt oblique co-ordinates, for a term in $y_1' y_2'$ then appears, nor displace the origin, for first-order terms are then introduced. It follows that it will suffice to consider a rotation of the x_1-x_2 planes about the ray axis. We may therefore write

$$x_1 + \mathrm{i}x_2 = (y_1 + \mathrm{i}y_2)\,e^{\mathrm{i}\chi}, \tag{3.3.7}$$

where $\chi(z)$ is real.

It would be possible, with the help of the partial integration rule, to follow the relations between the various ray variables which have been and will be introduced but, in fact, one is interested in relating ray variables to the ray direction only when considering rays in regions of *field-free* space. If we then choose (x_i, z) to be rectangular co-ordinates,

$$p_{xi} = \mathsf{p}x_i' \tag{3.3.8}$$

since all but the first term of (3.3.3) vanish, so that p_{xi} gives the ray direction directly.‡ Moreover, the ray variables will transform

‡ Since in this section we consider the transformation from x_i to other co-ordinates, it is advisable to write p_i more explicitly as p_{xi}.

exactly as the off-axis co-ordinates; for instance, (3.3.7) would give the relation

$$p_{x1}+ip_{x2}=(p_{y1}+ip_{y2})e^{i\chi}. \tag{3.3.9}$$

Now we find, from the partial integration rule, that (3.3.3) may be transformed into (3.3.4) by means of (3.3.7) if and only if p is identical in both formulae,

$$\mathsf{t}=\mathsf{p}\chi' \tag{3.3.10}$$

and

$$\left.\begin{aligned}\mathsf{e}_1&=\mathsf{p}\chi'^2-\tfrac{1}{2}(\mathsf{Y}_1+\mathsf{Y}_2)-\tfrac{1}{2}(\mathsf{Y}_1-\mathsf{Y}_2)\cos 2\chi,\\ \mathsf{e}_2&=\mathsf{p}\chi'^2-\tfrac{1}{2}(\mathsf{Y}_1+\mathsf{Y}_2)+\tfrac{1}{2}(\mathsf{Y}_1-\mathsf{Y}_2)\cos 2\chi,\\ \mathsf{f}&=-\tfrac{1}{2}(\mathsf{Y}_1-\mathsf{Y}_2)\sin 2\chi.\end{aligned}\right\} \tag{3.3.11}$$

Such a transformation exists, therefore, only if we can find three functions $\mathsf{Y}_1(z)$, $\mathsf{Y}_2(z)$ and $\chi(z)$ to satisfy the four equations (3.3.10) and (3.3.11). The appropriate condition is obtained by deriving from (3.3.11) the relation

$$\tan 2\chi=2\mathsf{f}/(\mathsf{e}_1-\mathsf{e}_2), \tag{3.3.12}$$

and noting that this is compatible with (3.3.10) only if

$$\mathsf{p}\{\mathsf{f}'(\mathsf{e}_1-\mathsf{e}_2)-\mathsf{f}(\mathsf{e}_1'-\mathsf{e}_2')\}=\mathsf{t}\{(\mathsf{e}_1-\mathsf{e}_2)^2+4\mathsf{f}^2\}. \tag{3.3.13}$$

The relation (3.3.13) should therefore be named the *orthogonality condition*; we shall consider only systems for which this condition is satisfied. The axes (y_i, z) for which the paraxial variational function assumes its orthogonal form will be termed the *principal axes*, and the surfaces whose equations are $y_1=0$ and $y_2=0$ will be referred to as the *principal sections* of the system.

It is not difficult to see that (3.3.13) must be satisfied for systems of rotational symmetry for, in order that (3.3.3) should be rotationally symmetrical, we must have $\mathsf{e}_1=\mathsf{e}_2$ and $\mathsf{f}=0$. In Chapter 5 we shall consider systems of 'mirror symmetry' for which m is even in one of the off-axis co-ordinates, say x_2. The condition (3.3.13) is again satisfied, for in this case $\mathsf{t}=\mathsf{f}=0$.

If (3.3.13) is satisfied, (3.3.10) and (3.3.11) yield

$$\left.\begin{aligned}\mathsf{Y}_1&=\mathsf{p}^{-1}\mathsf{t}^2-\tfrac{1}{2}(\mathsf{e}_1+\mathsf{e}_2)+\tfrac{1}{2}\sqrt{((\mathsf{e}_1-\mathsf{e}_2)^2+4\mathsf{f}^2)},\\ \mathsf{Y}_2&=\mathsf{p}^{-1}\mathsf{t}^2-\tfrac{1}{2}(\mathsf{e}_1+\mathsf{e}_2)-\tfrac{1}{2}\sqrt{((\mathsf{e}_1-\mathsf{e}_2)^2+4\mathsf{f}^2)}.\end{aligned}\right\} \tag{3.3.14}$$

The special case that $\mathsf{Y}_1(z)=\mathsf{Y}_2(z)$ is very important, and we shall then say that the paraxial imaging is *Gaussian*. Obviously, this condition can be satisfied only if $\mathsf{e}_1=\mathsf{e}_2$ and $\mathsf{f}=0$, which is precisely the requirement which is to be satisfied if the variational function is to be rotationally symmetrical. It also follows from the fact that (3.3.12) breaks down, leaving $\chi(z)$ determined only by (3.3.10) and hence arbitrary to the extent of an additive constant, that *a Gaussian paraxial system is one whose variational function is rotationally symmetrical about the ray axis.*

The ray equations (3.3.6) contain a first-order term, but we have already seen in § 1.4 how this may be eliminated. We may obtain the ray equations in a form better suited for numerical integration by writing

$$y_i = \mathsf{P}^{-\frac{1}{2}} q_i. \tag{3.3.15}$$

We then find, with the help of the partial integration rule, that the transform of (3.3.4) may be written as

$$m^{(2)} = \tfrac{1}{2}(q_1'^2 - \mathsf{Q}_1 q_1^2) - \tfrac{1}{2}(q_2'^2 - \mathsf{Q}_2 q_2^2), \tag{3.3.16}$$

where
$$\mathsf{Q}_i = \mathsf{P}^{-1}\mathsf{Y}_i + \tfrac{1}{4}(\mathsf{P}'/\mathsf{P})^2 - \tfrac{1}{2}(\mathsf{P}''/\mathsf{P}). \tag{3.3.17}$$

The ray equations derivable from (3.3.16) are

$$\frac{d^2 q_1}{dz^2} + \mathsf{Q}_1 q_1 = 0 \quad \text{and} \quad \frac{d^2 q_2}{dz^2} + \mathsf{Q}_2 q_2 = 0, \tag{3.3.18}$$

and the paraxial approximation to the ray variables is given by the simple formula

$$p_{qi} = q_i'. \tag{3.3.19}$$

On considering the form of the co-ordinate system adopted and on noting the relation (3.3.7), one might think it advantageous to combine the two off-axis co-ordinates into one complex co-ordinate. In fact, a complex notation tends to simplify calculations when the paraxial system is Gaussian but to complicate them otherwise. Let us therefore conclude the section by listing the co-ordinates and transformations which are useful in the discussion of Gaussian systems.

We may write

$$\mathsf{e} = \mathsf{e}_1 = \mathsf{e}_2, \quad \mathsf{V} = \mathsf{Y}_1 = \mathsf{Y}_2, \quad \mathsf{W} = \mathsf{Q}_1 = \mathsf{Q}_2 \tag{3.3.20}$$

and combine the off-axis co-ordinates and ray variables as follows:

$$u = x_1 + \mathrm{i}x_2, \quad v = y_1 + \mathrm{i}y_2, \quad w = q_1 + \mathrm{i}q_2, \tag{3.3.21}$$

$$p_u = p_{x1} + \mathrm{i}p_{x2}, \quad p_v = p_{y1} + \mathrm{i}p_{y2}, \quad p_w = p_{q1} + \mathrm{i}p_{q2}. \tag{3.3.22}$$

The transformations are

$$u = v\,e^{\mathrm{i}\chi}, \quad v = \mathsf{P}^{-\frac{1}{2}} w, \tag{3.3.23}$$

where χ is defined by (3.3.10). The various forms of the variational function are

$$m^{(2)} = \tfrac{1}{2}\{\mathsf{p}\bar{u}'u' - \mathrm{i}\mathsf{t}(\bar{u}'u - \bar{u}u') + \mathsf{e}\bar{u}u\}, \tag{3.3.24}$$

$$m^{(2)} = \tfrac{1}{2}(\mathsf{p}\bar{v}'v' - \mathsf{V}\bar{v}v), \tag{3.3.25}$$

and

$$m^{(2)} = \tfrac{1}{2}(\bar{w}'w' - \mathsf{W}\bar{w}w), \tag{3.3.26}$$

and the ray equations corresponding to the second and third of these forms are

$$\frac{d}{dz}\left(\mathsf{p}\frac{dv}{dz}\right) + \mathsf{V}v = 0 \tag{3.3.27}$$

and

$$\frac{d^2w}{dz^2} + \mathsf{W}w = 0. \tag{3.3.28}$$

The relations (3.3.14) and (3.3.17) now become

$$\mathsf{V} = \mathsf{p}^{-1}\mathsf{t}^2 - \mathsf{e} \tag{3.3.29}$$

and

$$\mathsf{W} = \mathsf{p}^{-1}\mathsf{V} + \tfrac{1}{4}(\mathsf{p}'/\mathsf{p})^2 - \tfrac{1}{2}(\mathsf{p}''/\mathsf{p}). \tag{3.3.30}$$

It is easy to verify that the variational function (1.5.9) may be expressed in the form (3.3.24) by taking $\mathsf{t} = \tfrac{1}{2}\mathsf{H}$ and $\mathsf{e} = 0$. Equations (3.3.27) and (3.3.29) then give immediately the paraxial ray equation (1.5.13).

Let us now compare the behaviour of rays determined by the ray equations (3.3.6), which hold for any paraxial system satisfying the orthogonality condition, with that of rays determined by (3.3.27), which holds only for Gaussian systems such as paraxial systems of rotational symmetry. As we saw in §§ 1.4 and 1.5, the general solution of (3.3.27) may be written as

$$v(z) = v_o g(z) + v_a h(z), \tag{3.3.31}$$

where $g(z)$ and $h(z)$ are solutions of (3.3.27) satisfying the boundary conditions

$$\left.\begin{aligned} g_o &= 1, & g_a &= 0, \\ h_o &= 0, & h_a &= 1. \end{aligned}\right\} \tag{3.3.32}$$

The planes $z = z_o$ and $z = z_a$ are supposed to be the object and aperture planes; these must be non-conjugate so that (3.3.32) can be satisfied. If the object plane is imaged upon the image plane $z = z_b$, then

$$h_b = 0 \tag{3.3.33}$$

and the magnification is given by

$$M = g_b. \tag{3.3.34}$$

If

$$k = \mathsf{p}(g'h - gh'), \tag{3.3.35}$$

then k is independent of z so that

$$k=-\mathrm{p}_o h_o', \quad k=\mathrm{p}_a g_a' \quad \text{and} \quad k=-\mathrm{p}_b g_b h_b'. \tag{3.3.36}$$

We see from (3.3.5), (3.3.21), (3.3.22) and (3.3.31) that k may be written in the form

$$k=\frac{\partial p_v}{\partial v_o}\frac{\partial v}{\partial v_a}-\frac{\partial p_v}{\partial v_a}\frac{\partial v}{\partial v_o}, \tag{3.3.37}$$

and hence, on comparison with (2.3.3), that k is a form of the Lagrange invariant.

The solution of the equations (3.3.6) may be written as

$$y_i(z)=y_{io}g_i(z)+y_{ia}h_i(z) \quad (i \text{ n.t.b.s.}). \tag{3.3.38}$$

The pairs of functions $g_1(z)$, $h_1(z)$ and $g_2(z)$, $h_2(z)$ satisfy the first and second of equations (3.3.6), respectively, together with the boundary conditions (3.3.32). Moreover, each pair determines a constant k_1 or k_2 by means of (3.3.35); the relations (3.3.36) still hold if g, h, k are replaced by g_1, h_1, k_1 or g_2, h_2, k_2. The single condition (3.3.33) for the planes $z=z_o$ and $z=z_b$ to be conjugate is now replaced by the pair of conditions

$$h_{ib}=0, \tag{3.3.39}$$

and in place of $v_b=Mv_o$ we have the relations

$$y_{ib}=M_i y_{io} \quad (i \text{ n.t.b.s.}), \tag{3.3.40}$$

where

$$M_i=g_{ib}, \tag{3.3.41}$$

which clearly represents an anisotropic magnification unless $M_1=M_2$. If both the conditions (3.3.39) are satisfied so that a point is imaged as a point, the imaging is said to be *stigmatic*, but if it is impossible for both conditions to be satisfied simultaneously (i.e. for the same choice of z_b), the imaging is termed *astigmatic*. In the latter case, a point object is imaged as a line in either of the planes $z=z_{b1}$, $z=z_{b2}$, where $h_1(z_{b1})=0$, $h_2(z_{b2})=0$. The separation between the planes, $|z_{b1}-z_{b2}|$, is known as the *astigmatic difference*.

We see that, provided the orthogonality condition is satisfied, any electron-optical system may display the same paraxial imaging properties as a system of rotational symmetry. There are, however, the differences that the pair of imaging conditions (3.3.39) must be satisfied instead of the single condition (3.3.33), and that a further condition is to be satisfied if the magnification is to be isotropic. If we consider only the projections of rays upon a principal section, there is no difference between Gaussian and orthogonal systems.

If we consider (3.3.18) instead of (3.3.6), and (3.3.28) instead of (3.3.27), the above results are unaltered except that y_i is replaced by q_i, v by w, and p by unity, and the magnification M_i would not be given by (3.3.41) but by

$$M_i = \mathsf{p}_o^{\frac{1}{2}} \mathsf{p}_b^{-\frac{1}{2}} g_{ib}. \tag{3.3.41a}$$

There is no difficulty in modifying the above formulae for the consideration of semi-telescopic or telescopic imaging instead of stigmatic imaging, but it is not proposed that we should throughout consider all these cases explicitly; it will suffice to indicate the modification to be made.

If we are interested in the formation of an image effectively at infinity, we should assume $z = z_b$ to be any plane conveniently located in the field-free image space, where the co-ordinates will be rectangular. Then, in terms of the Gaussian system, the complex ray variable p_{vb} will be linearly related to the object co-ordinate v_o by $p_{vb} = Mv_o$, where $M = \mathsf{p}_b g_b'$, if the imaging condition $h_b' = 0$ is satisfied. It is interesting to note from (1.4.21) and (1.4.22) that if the aperture plane lies in the field-free image space, so that we may take $z_b = z_a$,

$$k = -p_o/f_o, \tag{3.3.42}$$

where f_o is the first, i.e. 'object side', focal length.

If the object plane is at infinity, we should similarly replace v_o in (3.3.31) by p_{vo}, the boundary conditions (3.3.32) being modified to read

$$\left.\begin{aligned} g_o' &= \mathsf{p}_o^{-1}, & g_a &= 0, \\ h_o' &= 0, & h_a &= 1, \end{aligned}\right\} \tag{3.3.43}$$

with corresponding modification of (3.3.36) to read $k = h_o$. The same rules hold for orthogonal systems with appropriate modifications.

3.4. Gaussian dioptrics

Let us now consider, to the paraxial approximation, the optical coupling between two field-free regions. We may set up Cartesian co-ordinates in each region which coincide with the principal axes of the electron-optical system. If we confine our attention to the projections of rays upon a principal section, there will be no need to distinguish between Gaussian and orthogonal systems. Let us drop the suffix i and denote by x the off-axis co-ordinate measured along one of the principal axes and by p_x the corresponding ray variable which will be given by $p_x = px'$, since p is a constant in each of the regions considered. The regions will be referred to as the 'object' and 'image' spaces and characterized by the suffixes o and b, respectively.

We may define two solutions $s(z)$ and $t(z)$ of the appropriate ray equation (3.3.6) which satisfy the boundary conditions

$$s_o = 1, \quad s'_o = 0, \quad t_b = 1, \quad t'_b = 0. \tag{3.4.1}$$

These solutions will be distinct if the coupling is assumed not to be telescopic. Since $s(z)$ and $t(z)$ are linear in the two spaces, we may write

$$s(z_b) = -(z_b - z_{fb})/f_b, \quad t(z_o) = (z_o - z_{fo})/f_o \tag{3.4.2}$$

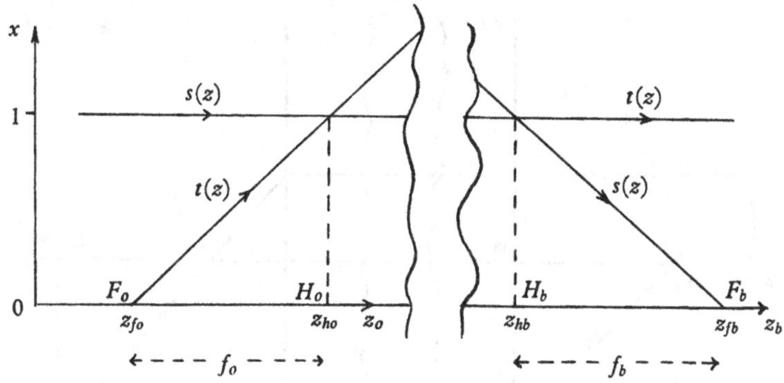

Fig. 3.2. The focal and principal points.

for their behaviour in the image and object spaces, respectively (fig. 3.2). The points $z_o = z_{fo}$ and $z_b = z_{fb}$ upon the ray axis are known as *the first and second focal points*, and f_o and f_b as *the first and second focal lengths*. We obtain at once from the Lagrange invariant $p(s't - st')$ that

$$f_o/f_b = p_o/p_b. \tag{3.4.3}$$

If we now consider the rays defined by the function $\alpha s(z) + \beta t(z)$, where α and β are arbitrary parameters, we see that the off-axis co-ordinates in the planes with co-ordinates z_o and z_b are given by

$$\left.\begin{aligned} x_o &= \alpha + \{(z_o - z_{fo})/f_o\}\beta, \\ x_b &= -\{(z_b - z_{fb})/f_b\}\alpha + \beta . \end{aligned}\right\} \tag{3.4.4}$$

The points (z_o, x_o) and (z_b, x_b) can be connected by a manifold of rays only if the determinant of (3.4.4) vanishes, i.e. if

$$(z_o - z_{fo})(z_b - z_{fb}) + f_o f_b = 0. \tag{3.4.5}$$

This is *Newton's relation*. The magnification is given by

$$M = f_o/(z_o - z_{fo}) \quad \text{or} \quad M = -(z_b - z_{fb})/f_b. \tag{3.4.6}$$

The imaging is said to be *real* if the conjugate points lie within the physical limits of the coupled spaces, and *virtual* otherwise.

The relations (3.4.5) may be rearranged as

$$\frac{f_o}{z_{ho} - z_o} + \frac{f_b}{z_b - z_{hb}} = 1, \tag{3.4.7}$$

where the denominators may be interpreted as the distances along the ray axis, measured *away* from the imaging system, of the conjugate points from the planes $z_o = z_{ho}$ and $z_b = z_{hb}$, where

$$z_{ho} = z_{fo} + f_o, \quad z_{hb} = z_{fb} - f_b. \tag{3.4.8}$$

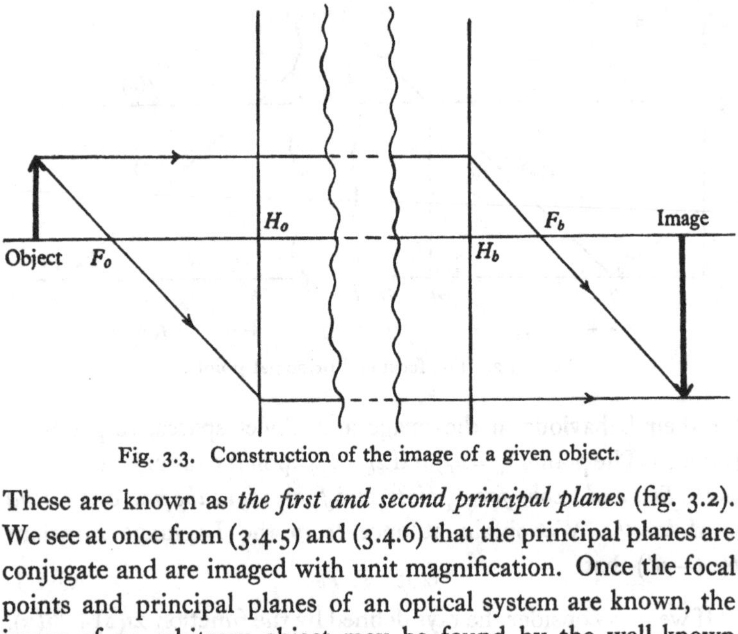

Fig. 3.3. Construction of the image of a given object.

These are known as *the first and second principal planes* (fig. 3.2). We see at once from (3.4.5) and (3.4.6) that the principal planes are conjugate and are imaged with unit magnification. Once the focal points and principal planes of an optical system are known, the image of an arbitrary object may be found by the well-known construction indicated in fig. 3.3.

It is worth remarking that in most electron-optical instruments the principal planes are 'crossed' in the sense that the z-coordinate of the second principal plane is smaller than that of the first principal plane.‡ It follows that both principal planes cannot lie

‡ See ref. (1), p. 431. Note that this rule does not apply to the 'osculating cardinal points' which are to be introduced later in this section (cf. P. A. Sturrock, *C.R. Acad. Sci., Paris*, **233** (1951), 401–3).

in their appropriate regions of field-free space; this point has been brought out in fig. 1.3.

Let us suppose that $(z_o, 0)$ and $(z_b, 0)$ are a pair of conjugate points and that $(z_o+\delta z_o, 0)$ and $(z_b+\delta z_b, 0)$ are another such pair. Then we find from (3.4.5), with the help of (3.4.6) and (3.4.3), that

$$\delta z_b/\delta z_o = (p_b/p_0)\, M^2, \tag{3.4.9}$$

a result which has already been established as (2.5.26). If

$$(z_o+\delta z_o, \delta x_o) \quad \text{and} \quad (z_b+\delta z_b, \delta x_b)$$

are another pair of conjugate points, the ray through $(z_o, 0)$ and $(z_o+\delta z_o, \delta x_o)$ must pass through $(z_b, 0)$ and $(z_b+\delta z_b, \delta x_b)$. Writing ω for dx/dz and noting that $\delta x_b = M\delta x_o$ to zero order in δz_o, we obtain

$$p_0\,\omega_o = Mp_b\,\omega_b, \tag{3.4.10}$$

which is the *Lagrange relation* established in § 1.4. We see from (3.4.3), (3.4.6) and (3.4.10) that $\omega_o=\omega_b$ if $z_o=z_{fo}+f_b$ and $z_b=z_{fb}-f_o$; these co-ordinates specify the *nodal points*. We may say that rays connecting these points have unit angular magnification. It should be noted that if $p_0=p_b$ the nodal points coincide with the principal points, i.e. the points where the ray axis intersects the principal planes. The focal points, the nodal points and the principal points are collectively named the *cardinal points* of a system.

We have assumed in this section that the object and image spaces are field-free and that they are at a finite distance from the refracting field. These assumptions are strictly incompatible, for electromagnetic fields extend indefinitely unless screened. However, if the field falls off fast enough, the rules which we have established may be satisfied by the *asymptotes* to the rays. The relevant criteria will be established in the next section.

Although in light optics the object and image are almost always located in regions where the refractive index is homogeneous and isotropic, in electron optics the object or image—or both—may be located where the refractive field is strong. The preceding treatment of Gaussian dioptrics then breaks down, since rays in the object and image spaces are not straight. However, it is possible to introduce *osculating* sets of cardinal points, as they have been named by Glaser,‡ which must be defined with reference to a pair of planes,

‡ Ref. (24).

say $z=z_o$ and $z=z_b$, by considering instead of the real rays *fictitious* rays which are defined only in the object and image spaces and there coincide with the tangents to the real rays where they cross the planes of reference; one will usually choose a *conjugate* pair of reference planes. These fictitious rays are the rays of a possible optical system, since their Lagrange invariant is the same as that of their associated real rays and is therefore an invariant of the fictitious optical transformation, provided only that we assign to p_o and p_b in the fictitious system the values p_o and p_b.

Since the real and fictitious rays are indistinguishable provided we work only to the first order in the increments δz_o and δz_b, the relations (3.4.3), (3.4.9) and (3.4.10) will hold for the planes of reference, and the relations (3.4.5), (3.4.6) and (3.4.7) also will remain valid provided that we replace z_o and z_b by $z_o+\delta z_o$ and $z_b+\delta z_b$ and work to the first order only in the increments. Thus we may state that whereas the cardinal points characterize the optical connexion between the finite object and image spaces when these spaces are field-free, the osculating cardinal points characterize only the connexion between the infinitesimal regions in the neighbourhoods of the reference planes if these planes are located in the refracting field.

Suppose that the co-ordinates of conjugate planes are written as $z_o(t)$ and $z_b(t)$ and their magnification is written as $M(t)$, where the parameter t enumerates the pairs of planes. The osculating cardinal points associated with a given pair of conjugate planes may then be determined by ensuring that the relations (3.4.6) are satisfied, with the same f_o, f_b, z_{fo} and z_{fb}, not only for t but for $t+\delta t$. Hence we obtain the formulae

$$f_o=-\frac{M^2\,\mathrm{d}z_o/\mathrm{d}t}{\mathrm{d}M/\mathrm{d}t}, \qquad f_b=-\frac{\mathrm{d}z_b/\mathrm{d}t}{\mathrm{d}M/\mathrm{d}t}, \tag{3.4.11}$$

$$z_{fo}=z_o+\frac{M\,\mathrm{d}z_o/\mathrm{d}t}{\mathrm{d}M/\mathrm{d}t}, \qquad z_{fb}=z_b-\frac{M\,\mathrm{d}z_b/\mathrm{d}t}{\mathrm{d}M/\mathrm{d}t}, \tag{3.4.12}$$

$$z_{ho}=z_o-\frac{M(M-1)\,\mathrm{d}z_o/\mathrm{d}t}{\mathrm{d}M/\mathrm{d}t}, \quad z_{hb}=z_b-\frac{(M-1)\,\mathrm{d}z_b/\mathrm{d}t}{\mathrm{d}M/\mathrm{d}t}, \tag{3.4.13}$$

for the osculating focal lengths and co-ordinates of the osculating cardinal points.

One should expect the osculating cardinal points to vary in position with the reference planes, but there exists a class of fields, which Glaser has termed 'Newtonian', for which the relation (3.4.5) is satisfied for all pairs of conjugate planes with the same values of f_o, f_b, z_{fo} and z_{fb}, although the relations (3.4.6) are not necessarily satisfied.‡ The most important member of this class is the bell-shaped magnetic field, originally studied by Glaser,§ of the form

$$\mathsf{H}(z) = \mathsf{H}_0(1+(z/a)^2)^{-1}. \tag{3.4.14}$$

The paraxial ray equation (1.5.13) now becomes

$$\frac{d^2v}{d(z/a)^2} + \frac{\frac{1}{4}a^2p^{-2}\mathsf{H}_0^2}{(1+(z/a)^2)^2}v = 0, \tag{3.4.15}$$

which may be solved analytically by the transformation

$$z = a\cot\phi \quad (0<\phi<\pi). \tag{3.4.16}$$

The general solution is found to be

$$v(\phi) = K\operatorname{cosec}\phi\,\sin(\omega\phi+\kappa), \tag{3.4.17}$$

where K and κ are parameters and

$$\omega^2 = 1+\tfrac{1}{4}a^2p^{-2}\mathsf{H}_0^2. \tag{3.4.18}$$

It follows at once from (3.4.17) that ϕ_o and ϕ_b correspond to the co-ordinates of conjugate planes if the relation

$$\phi_o - \phi_b = \pi/\omega \tag{3.4.19}$$

is satisfied and that the magnification is given by

$$M = -\sin\phi_o/\sin\phi_b. \tag{3.4.20}$$

There is no difficulty in evaluating the formulae (3.4.11), (3.4.12), and (3.4.13) if t is taken as ϕ_o and z_o, z_b and M are expressed as functions of this parameter by means of (3.4.16), (3.4.19) and (3.4.20); we find that

$$\left.\begin{aligned} f_o &= a\operatorname{cosec}\pi/\omega, & z_{fo} &= a\cot\pi/\omega, & z_{ho} &= a\cot\pi/2\omega,\\ f_b &= a\operatorname{cosec}\pi/\omega, & z_{fb} &= -a\cot\pi/\omega, & z_{hb} &= -a\cot\pi/2\omega, \end{aligned}\right\} \tag{3.4.21}$$

so that the osculating cardinal points have the same positions for all pairs of conjugate planes. In other words, the Gaussian imaging properties of this field may be characterized by a fixed set of cardinal points even though the object and image cannot be in field-free space.

‡ W. Glaser and E. Lammel, *Ann. Phys., Lpz.*, **40** (1941), 367–84; R. G. E. Hutter, ref. (23); P. Funk, *Acta Phys. Austriaca*, **4** (1950), 304–8.

§ W. Glaser, *Z. Phys.* **117** (1941), 285–315.

Since, to the paraxial approximation, any optical system is characterized by its cardinal points, it must be possible to obtain the cardinal points of a compound optical system from the cardinal points of its elements. It is not difficult to establish an appropriate geometrical construction, but we may obtain a much neater solution of the problem by means of a matrix notation.

Let us consider once more the reference planes $z = z_o$ and $z = z_b$, which may be situated in the refractive field, but discard the condition that they should be conjugate; the co-ordinates p_y and y will now be more appropriate than p_x and x. There must be a linear relationship between the values of the canonical variables in the two planes so that we may write, in matrix notation,

$$\begin{pmatrix} p_{yb} \\ y_b \end{pmatrix} = \begin{pmatrix} A & B \\ C & D \end{pmatrix} \begin{pmatrix} p_{yo} \\ y_o \end{pmatrix}. \tag{3.4.22}$$

The optical coupling between the two planes is then characterized by the square matrix appearing in (3.4.22). It is seen at once that the matrix characterizing a compound system will be simply the product of the matrices which characterize its elements. To complete the solution, it is only necessary to relate the elements of the matrix in (3.4.22) to the cardinal points—or the osculating cardinal points—of the system.

However, let us first note one or two interesting properties of the matrices. It is not to be expected that every two-by-two matrix should represent an optical transformation. The appropriate condition which must be satisfied is readily found, by consideration of the Lagrange invariant, to be that

$$AD - BC = 1, \tag{3.4.23}$$

i.e. that the determinant of the matrix has the value unity. On inverting the matrix, we now obtain the converse relation

$$\begin{pmatrix} p_{yo} \\ y_o \end{pmatrix} = \begin{pmatrix} D & -B \\ -C & A \end{pmatrix} \begin{pmatrix} p_{yb} \\ y_b \end{pmatrix}. \tag{3.4.24}$$

We may deduce from (3.4.22) that the condition that there should be stigmatic imaging of the planes is that $C = 0$ and that the magnification is then equal to D; there is semi-telescopic imaging if either A or D vanishes and the 'magnification' is then given by B or C, respectively; similarly, there is telescopic imaging, with 'magnification' A, if B vanishes.

On remarking that $p_y = py'$ and on constructing the rays $s(z)$ and $t(z)$ introduced at the beginning of this section, we find that the focal lengths and the co-ordinates of the focal points and principal planes are given by

$$f_o = -p_o(1/B), \qquad f_b = -p_b(1/B), \tag{3.4.25}$$

$$z_{fo} = z_o + p_o(A/B), \qquad z_{fb} = z_b - p_b(D/B), \tag{3.4.26}$$

and

$$z_{ho} = z_o + p_o(A-1)/B, \quad z_{hb} = z_b - p_b(D-1)/B. \tag{3.4.27}$$

Conversely, one may calculate the matrix elements from knowledge of the cardinal points and (3.4.23).

An interesting simple application of the Gaussian matrices is the calculation of the cardinal points of electrostatic lenses, of the type shown in fig. 1.1, to an approximation which is equivalent to replacing all electrodes by thin diaphragms with holes of very small radius. The axial field distribution may then be divided into sections in each of which the potential rises or falls linearly (so that the electric field is uniform) and at the junctions of which the derivative of the potential changes discontinuously. If we write down the matrices which characterize the various possible elements of such lenses, we shall be able to combine them as necessary in calculating the properties of a given lens.

The matrix M_{ue} which characterizes a length of uniform electric field may be found by solving (1.4.12) when $\Phi(z)$ is a linear function of z. To the non-relativistic approximation,

$$M_{ue} = \begin{pmatrix} 1 & 0 \\ 2\dfrac{(z_b - z_o)}{(p_o + p_b)} & 1 \end{pmatrix}. \tag{3.4.28}$$

The matrix M_{de} which characterizes a sudden increase $\Delta\Phi'$ in the axial field strength may be deduced immediately from (1.4.25); if both planes of reference coincide with the plane at which the discontinuity occurs, then to the non-relativistic approximation,

$$M_{de} = \begin{pmatrix} 1 & -\frac{1}{2}p^{-1}\Delta\Phi' \\ 0 & 1 \end{pmatrix}. \tag{3.4.29}$$

If there is a foil or fine grid in the lens, there will be a discontinuity in the field strength but no corresponding deflexion of the rays; the matrix characterizing a foil is therefore the unit matrix which may be ignored.

As an example, let us consider the 'bipotential' lens whose potential distribution is shown in fig. 3.4. Such a distribution would be produced by a pair of electrodes at $z=-\frac{1}{2}d$ and $z=\frac{1}{2}d$ held at potentials Φ_1 and Φ_2, respectively, if the electrodes were very thin and their openings very small. If M_1, M_2 and M_3 are the matrices characterizing the discontinuity at $z=-\frac{1}{2}d$, the uniform field between $z=-\frac{1}{2}d$ and $z=\frac{1}{2}d$, and the discontinuity at $z=\frac{1}{2}d$, respectively, the matrix for the lens is given by

$$M=M_3M_2M_1, \tag{3.4.30}$$

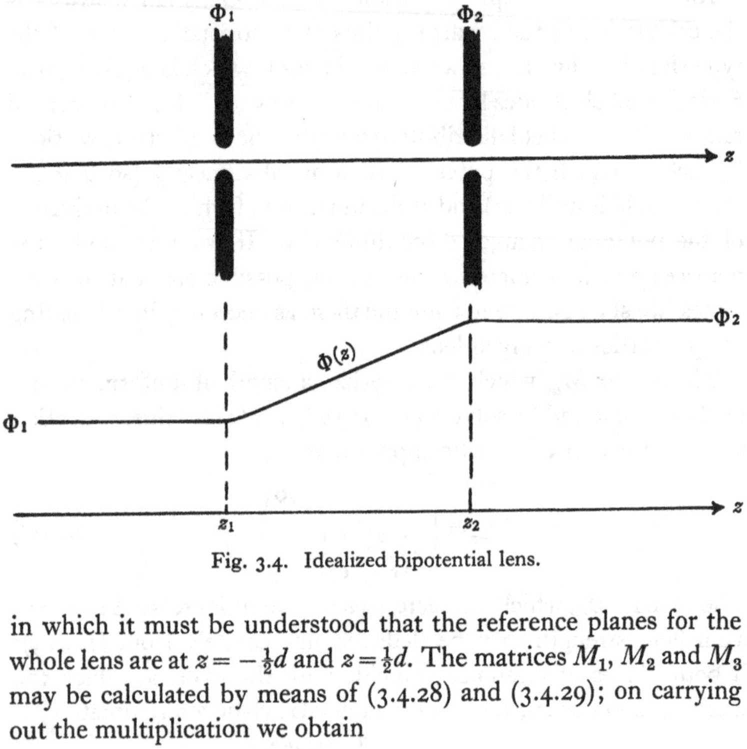

Fig. 3.4. Idealized bipotential lens.

in which it must be understood that the reference planes for the whole lens are at $z=-\frac{1}{2}d$ and $z=\frac{1}{2}d$. The matrices M_1, M_2 and M_3 may be calculated by means of (3.4.28) and (3.4.29); on carrying out the multiplication we obtain

$$M=\begin{pmatrix} \dfrac{3p_2-p_1}{2p_2} & -\dfrac{3}{4d}\dfrac{p_2-p_1}{p_1p_2}(\Phi_2-\Phi_1) \\ \dfrac{2d}{p_1-p_2} & \dfrac{3p_1-p_2}{2p_1} \end{pmatrix}. \tag{3.4.31}$$

The formulae (3.4.25) and (3.4.27) now lead to the following formula for the focal lengths and co-ordinates of the principal points of the bipotential lens:

$$\frac{1}{f_o}=\frac{3}{8d}\left(1-\frac{p_1}{p_2}\right)\left(\frac{\Phi_2}{\Phi_1}-1\right),\quad \frac{1}{f_b}=\frac{3}{8d}\left(\frac{p_2}{p_1}-1\right)\left(\frac{\Phi_2}{\Phi_1}-1\right), \qquad (3.4.32)$$

$$z_{ho}=-\tfrac{1}{2}d-\frac{4d}{3}\frac{\Phi_1}{\Phi_2-\Phi_1},\quad z_{hb}=\tfrac{1}{2}d-\frac{4d}{3}\frac{\Phi_2}{\Phi_2-\Phi_1}. \qquad (3.4.33)$$

We see from (3.4.33) that $z_{hb}-z_{ho}=-\frac{1}{3}d$, so that the principal planes are indeed 'crossed'; they are both situated outside the lens on the low potential side. We also see from (3.4.32) that the focal lengths are always positive, i.e. that the lens is always convergent.‡

In conclusion, we may consider the question whether, by a suitable choice of field, Gaussian dioptrics might be made to hold not merely to the paraxial approximation but for finite object and aperture. That the answer is negative may be seen by comparing the Lagrange relation (3.4.9) with the sine relation (2.5.19); these are compatible only to the paraxial approximation beyond which ω in (3.4.9) would be replaced by $\tan\omega$ if Gaussian dioptrics still remained valid.

3.5. Perturbation characteristic functions

It is now necessary to reconsider the properties of perturbation characteristic functions which were introduced in §2.4. Although the differences between the two treatments appear to be slight, the present formulation will prove to be better adapted to aberration calculation.

Let the variational function now be subject to a perturbation so that its perturbed form $\mathsf{P}m$ is given by

$$\mathsf{P}m=m+\varpi m^{\mathrm{I}}+\dots, \qquad (3.5.1)$$

where ϖ is the perturbation parameter. It is worth noticing that the rule established at the end of §2.4 now implies that it is permissible to add any multiple of m to m^{I}. We may introduce similar expansions for the perturbed forms of the off-axis co-ordinates and the ray variables:

$$\mathsf{P}x_i=x_i+\varpi x_i^{\mathrm{I}}+\dots, \qquad (3.5.2)$$

and

$$\mathsf{P}n_i=n_i+\varpi n_i^{\mathrm{I}}+\dots. \qquad (3.5.3)$$

‡ For further examples, see C. Fert, *J. Phys. Radium*, **13** (1952), 83A–90A.

However, we write, as before,

$$\mathsf{P}V_{ab} = V_{ab} + \varpi\, {}^*V^{\mathrm{I}}_{ab} + \ldots. \tag{3.5.4}$$

It will be recalled that the first-order contribution to the perturbed form of the point characteristic function is the sum of two terms: one will be due to the perturbation of the ray, the variational function being taken in its unperturbed form; and the other due to the perturbation of the variational function, the path of integration being the unperturbed ray. On noting that the former may be evaluated by means of (3.2.7), we see that

$$ {}^*V^{\mathrm{I}}_{ab} = p_{ib}x^{\mathrm{I}}_{ib} - p_{ia}x^{\mathrm{I}}_{ia} + V^{\mathrm{I}}_{ab}, \tag{3.5.5}$$

where

$$V^{\mathrm{I}}_{ab} = \int_{z_b}^{z_b} m^{\mathrm{I}} dz, \tag{3.5.6}$$

the integral being evaluated along the unperturbed ray. On taking the first-order contribution to the perturbed form of Hamilton's relation (3.2.7) and making use of (3.5.5), we obtain

$$\delta V^{\mathrm{I}}_{ab} = (p^{\mathrm{I}}_{ib}\,\delta x_{ib} - x^{\mathrm{I}}_{ib}\,\delta p_{ib}) - (p^{\mathrm{I}}_{ia}\,\delta x_{ia} - x^{\mathrm{I}}_{ia}\,\delta p_{ia}), \tag{3.5.7}$$

which is the new form of the first-order perturbation relation.

From a comparison of (3.5.7) and (2.4.10), it is obvious that a set of perturbation characteristic functions may be derived from the former which have the same properties as those derived from the latter except that the three-component vectors x_r and p_r are now replaced by the two-component vectors x_i and p_i. For instance, if the planes $z = z_a$ and $z = z_b$ are not stigmatically imaged, we may prescribe that

$$x^{\mathrm{I}}_{ia} = x^{\mathrm{I}}_{ib} = 0, \tag{3.5.8}$$

so that (3.5.7) reduces to

$$\delta V^{\mathrm{I}}_{ab} = p^{\mathrm{I}}_{ib}\,\delta x_{ib} - p^{\mathrm{I}}_{ia}\,\delta x_{ia}, \tag{3.5.9}$$

from which we see that V^{I}_{ab} may then be expressed as $V^{\mathrm{I}}_{ab}(x_{ia}, x_{ib})$, the 'point perturbation characteristic function', and that

$$p^{\mathrm{I}}_{ia} = -\frac{\partial V^{\mathrm{I}}_{ab}}{\partial x_{ia}}, \quad p^{\mathrm{I}}_{ib} = \frac{\partial V^{\mathrm{I}}_{ab}}{\partial x_{ib}}. \tag{3.5.10}$$

The functions introduced in § 2.4 all characterized the perturbation of the coupling between two spaces, but it would be convenient if we could introduce characteristic functions which give the ray perturbation in the image plane as a function of the ray co-ordinates in the object and aperture planes. This may be achieved as follows.

Let us introduce the object, aperture and image planes with co-ordinates z_o, z_a and z_b and suppose that the object and image planes are stigmatically imaged when the system is in its unperturbed state. Let us also introduce a 'current' plane $z = z_c$, where z_c can take arbitrary values. If we specify that the object and aperture planes should not be conjugate, we may prescribe that

$$x_{io}^{\mathrm{I}} = x_{ia}^{\mathrm{I}} = 0. \tag{3.5.11}$$

We shall now introduce a pair of ***aperture perturbation characteristic functions*** defined by

$$V_{oc}^{\mathrm{I}} = \int_{z_o}^{z_c} m^{\mathrm{I}} \mathrm{d}z, \quad V_{ac}^{\mathrm{I}} = \int_{z_a}^{z_c} m^{\mathrm{I}} \mathrm{d}z, \tag{3.5.12}$$

which may be expressed as $V_{oc}^{\mathrm{I}}(x_{io}, x_{ia})$ and $V_{ac}^{\mathrm{I}}(x_{io}, x_{ia})$, since these arguments determine rays uniquely.

If we now consider the first-order perturbation relations satisfied by these functions and bear in mind (3.5.11), we may establish the following set of simultaneous equations

$$\left.\begin{aligned} p_{ic}^{\mathrm{I}} \frac{\partial x_{ic}}{\partial x_{ja}} - x_{ic}^{\mathrm{I}} \frac{\partial p_{ic}}{\partial x_{ja}} &= \frac{\partial V_{oc}^{\mathrm{I}}}{\partial x_{ja}}, \\ p_{ic}^{\mathrm{I}} \frac{\partial x_{ic}}{\partial x_{jo}} - x_{ic}^{\mathrm{I}} \frac{\partial p_{ic}}{\partial x_{jo}} &= \frac{\partial V_{ac}^{\mathrm{I}}}{\partial x_{jo}}, \end{aligned}\right\} \tag{3.5.13}$$

where p_{ic} and x_{ic} are supposed expressed as functions of x_{io} and x_{ia}. The equations (3.5.13) may be solved for p_{ic}^{I} and x_{ic}^{I}.

The determinant Δ_c of the above set of equations is given by

$$\Delta = -\frac{\partial(p_1, x_1, p_2, x_2)}{\partial(x_{1o}, x_{2o}, x_{1a}, x_{2a})}, \tag{3.5.14}$$

where we have dropped the suffix c. This may be expanded as

$$\begin{aligned} \Delta = -\{x_{1o}, x_{2o}\}\{x_{1a}, x_{2a}\} + \{x_{1o}, x_{1a}\}\{x_{2o}, x_{2a}\} \\ -\{x_{1o}, x_{2a}\}\{x_{2o}, x_{1a}\}, \end{aligned} \tag{3.5.15}$$

so that, since each Lagrange bracket is constant, so must be Δ. It follows that Δ is given by either

$$\Delta = \frac{\partial(p_{1o}, p_{2o})}{\partial(x_{1a}, x_{2a})} \quad \text{or} \quad \Delta = \frac{\partial(p_{1a}, p_{2a})}{\partial(x_{1o}, x_{2o})}, \tag{3.5.16}$$

in which p_{io} and p_{ia} are supposed to have been expressed as functions of x_{io} and x_{ia}. It is obvious that Δ cannot vanish, for if the first, say, of the Jacobians were zero, it would be impossible for a ray to leave a point in the object plane with arbitrary direction.

Although, in the general case we have considered above, it is necessary to solve four simultaneous equations in order to obtain the ray perturbation in an arbitrary plane, the ray perturbation in the image plane may be obtained from only two simultaneous equations. If we note that

$$\partial x_{ib}/\partial x_{ja} = 0, \tag{3.5.17}$$

we see that the first pair of equations (3.5.13) may be solved to give

$$\left.\begin{aligned} x^{\mathrm{I}}_{1b} &= K^{-1}\left\{\frac{\partial p_{2b}}{\partial x_{1a}}\frac{\partial V^{\mathrm{I}}_{ob}}{\partial x_{2a}} - \frac{\partial p_{2b}}{\partial x_{2a}}\frac{\partial V^{\mathrm{I}}_{ob}}{\partial x_{1a}}\right\}, \\ x^{\mathrm{I}}_{2b} &= K^{-1}\left\{\frac{\partial p_{1b}}{\partial x_{2a}}\frac{\partial V^{\mathrm{I}}_{ob}}{\partial x_{1a}} - \frac{\partial p_{1b}}{\partial x_{1a}}\frac{\partial V^{\mathrm{I}}_{ob}}{\partial x_{2a}}\right\}, \end{aligned}\right\} \tag{3.5.18}$$

where

$$K = \partial(p_{1b}, p_{2b})/\partial(x_{1a}, x_{2a}). \tag{3.5.19}$$

It is important to note that the perturbation in the image plane may be obtained from the one characteristic function V^{I}_{ob}.

Let us now pass from the general case to the important special case that the unperturbed variational function is orthogonal. If we suppose that m is given in the form (3.3.4), we should replace p_i and x_i in the foregoing by p_{yi} and y_i. On using (3.3.5) and (3.3.38), we may easily solve (3.5.13) to obtain

$$y^{\mathrm{I}}_{ic} = k_i^{-1}\left\{g_{ic}\frac{\partial V^{\mathrm{I}}_{oc}}{\partial y_{ia}} - h_{ic}\frac{\partial V^{\mathrm{I}}_{ac}}{\partial y_{io}}\right\} \quad (i \text{ n.t.b.s.}). \tag{3.5.20}$$

We see from (3.3.39) that, in the image plane,

$$y^{\mathrm{I}}_{ib} = k_i^{-1} g_{ib}\frac{\partial V^{\mathrm{I}}_{ob}}{\partial y_{ia}} \quad (i \text{ n.t.b.s.}). \tag{3.5.21}$$

It is interesting to note from (3.5.21) that the ray perturbation can be 'referred back' to the object plane simply by omitting the term g_{ib}, since this factor represents the paraxial magnification of the system. It will therefore be possible to use formulae which hold for stigmatic imaging, with slight modification only, for the calculation of the aberrations of semi-telescopic systems, such as objective lenses, for which the image plane is effectively at infinity. If the object plane is at infinity, we should either introduce the modifications suggested at the end of § 3.3 or, preferably, interchange the roles of the object and image planes and proceed as suggested above.

The above formulae may, of course, be applied also if the unperturbed variational function is Gaussian; the suffix i may then be dropped from $g(z)$, $h(z)$ and k. However, as we saw in § 3.3, it is also possible to use complex co-ordinates. By adopting the expression (3.3.31) for the paraxial ray, we shall express the perturbation characteristic functions as functions of $\bar{v}_o$, v_o, $\bar{v}_a$ and v_a. On noting that

$$\delta y_i \frac{\partial}{\partial y_i} = \delta \bar{v} \frac{\partial}{\partial \bar{v}} + \delta v \frac{\partial}{\partial v}, \tag{3.5.22}$$

since each side represents the same operation, we see that the second of equations (3.3.21) leads to the relation

$$\frac{\partial}{\partial y_1} + \mathrm{i}\frac{\partial}{\partial y_2} = 2\frac{\partial}{\partial \bar{v}} \tag{3.5.23}$$

between the differential operators so that we may immediately write the complex forms of (3.5.20) and (3.5.21) as

$$v_c^{\mathrm{I}} = 2k^{-1}\left\{g_c \frac{\partial V_{oc}^{\mathrm{I}}}{\partial \bar{v}_a} - h_c \frac{\partial V_{ac}^{\mathrm{I}}}{\partial \bar{v}_o}\right\} \tag{3.5.24}$$

and

$$v_b^{\mathrm{I}} = 2k^{-1} g_b \frac{\partial V_{ob}^{\mathrm{I}}}{\partial \bar{v}_a}. \tag{3.5.25}$$

Since the formulae (3.5.20), (3.5.21), (3.5.24) and (3.5.25) do not involve $\mathsf{p}(z)$ they must hold in the same form if y_i is replaced by q_i and if v is replaced by w.

It has been assumed in this section that we are dealing with a real object such as that which is imaged by the objective lens of an electron microscope. However, it is possible for the image of an objective lens to be formed inside the refracting field of the projector lens which therefore forms an image not of a real object but of a *virtual object*, namely, the real image which *would* be formed by the objective lens if the projector lens were switched off. Let us determine what modifications should be made in perturbation calculations in such cases so that we shall know how to calculate the aberrations of projector lenses.

Let the object, aperture and image planes now be situated as in fig. 3.5; the additional plane $z = z_e$, the 'entrance' plane, will be assumed to be in field-free space. Since it is now necessary to distinguish between the real and virtual parts of rays, we shall define $\tilde{g}(z)$ and $\tilde{h}(z)$ to be the virtual portions of the fundamental rays $g(z)$ and $h(z)$. We see from fig. 3.5 that $\tilde{g}(z)$ and $\tilde{h}(z)$ will be linear, whereas $g(z)$ and $h(z)$ are solutions of the relevant paraxial ray equation, and that the appropriate boundary conditions are

$$\left.\begin{aligned}\tilde{g}_o=1,\quad g_a=0,\quad \tilde{g}_e=g_e,\quad \tilde{g}'_e=g'_e;\\ \tilde{h}_o=0,\quad h_a=1,\quad \tilde{h}_e=h_e,\quad \tilde{h}'_e=h'_e.\end{aligned}\right\}\tag{3.5.26}$$

It is also necessary to introduce, apart from m^{I} which characterizes the perturbation of the paraxial variational function, $\tilde{m}^{\mathrm{I}}$ which represents the corresponding perturbation of the virtual part of the system; thus $\tilde{m}^{\mathrm{I}}$ is the 'field-free' form of m^{I}.‡ The perturbation characteristic function for the complete system may now be obtained from the relation

$$V^{\mathrm{I}}_{ob}=\int_{z_o}^{z_e} m^{\mathrm{I}}\,dz+\int_{z_e}^{z_b} m^{\mathrm{I}}\,dz,\tag{3.5.27}$$

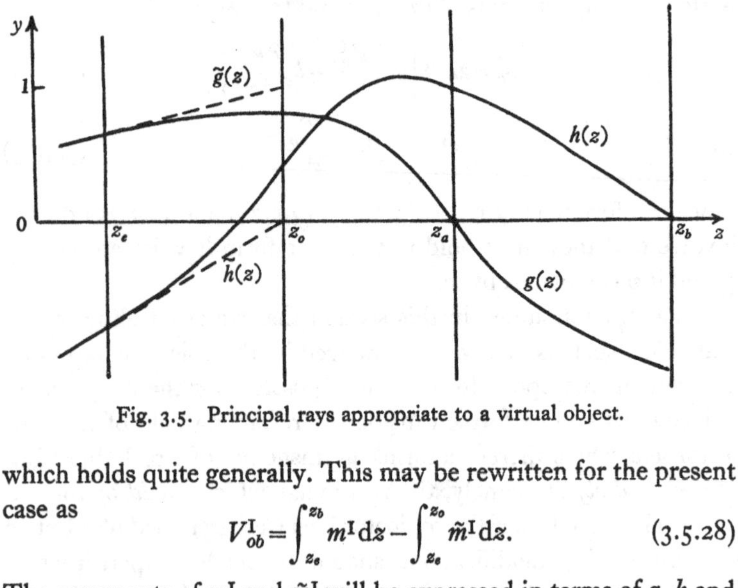

Fig. 3.5. Principal rays appropriate to a virtual object.

which holds quite generally. This may be rewritten for the present case as

$$V^{\mathrm{I}}_{ob}=\int_{z_e}^{z_b} m^{\mathrm{I}}\,dz-\int_{z_e}^{z_o} \tilde{m}^{\mathrm{I}}\,dz.\tag{3.5.28}$$

The arguments of m^{I} and $\tilde{m}^{\mathrm{I}}$ will be expressed in terms of g, h and $\tilde{g}$, $\tilde{h}$ respectively. It will be shown in § 5.6, where this formula will be applied, that it is convenient to regard a cathode-ray tube as forming an image of a virtual object.

Although the method of perturbation characteristic functions is now well established in electron optics—due largely to the publications of Glaser—aberrations are still often calculated by the method

‡ For instance, in §§ 4.2 and 4.3 $\tilde{m}^{(4)}$ would be given by $-\frac{1}{8}p\bar{u}'^2u'^2$ in (4.2.13) or by the term in $\mathscr{A}$ in (4.3.24).

of variation of parameters which was introduced into electron optics by Scherzer.‡ The latter is neither so neat nor so flexible as the former, but the two methods can be shown to be equivalent if we assume that the unperturbed variational function is Gaussian or orthogonal so that the method of variation of parameters can be applied. It will suffice to establish a typical formula, say (3.5.25), by this method.

We suppose that the variational function (3.3.25), whose solution is given by (3.3.31), is modified by the addition of ϖm^{I}, where $m^{\mathrm{I}} = m^{\mathrm{I}}(\bar{v}', v', \bar{v}, v, z)$ and ϖ is small, so that the ray is displaced to $v(z) + \varpi v^{\mathrm{I}}(z)$. We find, on working to the first order in ϖ, that $v^{\mathrm{I}}(z)$ is determined by the equation

$$\frac{\mathrm{d}}{\mathrm{d}z}\left\{\mathsf{P}\frac{\mathrm{d}v^{\mathrm{I}}}{\mathrm{d}z}\right\} - \mathsf{V}v^{\mathrm{I}} = -2\left[\frac{\mathrm{d}}{\mathrm{d}z}\left(\frac{\partial m^{\mathrm{I}}}{\partial \bar{v}'}\right) - \frac{\partial m^{\mathrm{I}}}{\partial \bar{v}}\right], \tag{3.5.29}$$

where we are to give v its unperturbed value in the function enclosed by square brackets. If we now write

$$v^{\mathrm{I}}(z) = \alpha(z)\,g(z) + \beta(z)\,h(z) \tag{3.5.30}$$

and specify that $h_b = 0$, we see that $v_b^{\mathrm{I}} = \alpha_b g_b$. We may substitute (3.5.30) for v^{I} in (3.5.29) and so solve for $\alpha(z)$ and $\beta(z)$, imposing the arbitrary but permissible condition that $\alpha' g + \beta' h = 0$. Hence we obtain

$$\alpha_b = -2k^{-1}\int_{z_0}^{z_b}\left[\frac{\mathrm{d}}{\mathrm{d}z}\left(\frac{\partial m^{\mathrm{I}}}{\partial \bar{v}'}\right) - \frac{\partial m^{\mathrm{I}}}{\partial \bar{v}}\right] h\,\mathrm{d}z, \tag{3.5.31}$$

where we have made use of (3.3.35). On integrating by parts, (3.5.31) becomes

$$\alpha_b = 2k^{-1}\int_{z_0}^{z_b}\left\{h'\frac{\partial}{\partial \bar{v}'} + h\frac{\partial}{\partial \bar{v}}\right\} m^{\mathrm{I}}\,\mathrm{d}z. \tag{3.5.32}$$

Now in the same way as we established (3.5.23) we may, by considering the transformation from variables $\bar{v}'$, v', $\bar{v}$, v to variables $\bar{v}_o$, v_o, $\bar{v}_a$, v_a determined by (3.3.31), show that

$$h'\frac{\partial}{\partial \bar{v}'} + h\frac{\partial}{\partial \bar{v}} = \frac{\partial}{\partial \bar{v}_a}. \tag{3.5.33}$$

We are now led to (3.5.25), where V_{ob}^{I} is defined by (3.5.6).

The fourth and fifth chapters will provide us with many examples of the application of aperture perturbation characteristic functions. Let us therefore conclude the section with a simple problem which

‡ See reference quoted on p. 24, n. ‡.

demonstrates the use of one of the other characteristic functions. It will be interesting to make use of the characteristic function U^{I}, since this has no counterpart in the theory of unperturbed systems.

Let us consider the projections of rays upon a principal section (z, y) and suppose that the refractive field is weak. If we work to the paraxial approximation, we may assume the variational function to be $m+m^{\mathrm{I}}$ (putting $\varpi=1$), where $m=\frac{1}{2}\mathsf{p}_0 y'^2$, p_0 being independent of z, and

$$m^{\mathrm{I}}=\tfrac{1}{2}\{(\mathsf{p}-\mathsf{p}_0)y'^2-\mathsf{Y}y^2\}. \tag{3.5.34}$$

It is easy to verify that the unperturbed or field-free system is described by a matrix equation (3.4.22) with $A=D=1$, $B=0$ and $C=\mathsf{p}_0^{-1}(z_b-z_o)$. The perturbed matrix may therefore be written as

$$\begin{pmatrix} A & B \\ C & D \end{pmatrix}=\begin{pmatrix} 1+A^{\mathrm{I}} & B^{\mathrm{I}} \\ \mathsf{p}_0^{-1}(z_b-z_o)+C^{\mathrm{I}} & 1+D^{\mathrm{I}} \end{pmatrix}. \tag{3.5.35}$$

We shall attempt to calculate the coefficients appearing in (3.5.35) by the methods of first-order perturbation theory; the approximation involved is the *thin-lens approximation* which we have already met in §§ 1.4 and 1.5.

We may deduce from (2.4.14) and (2.4.18) that if we express U^{I}_{ob}, defined by the integral (3.5.6), as $U^{\mathrm{I}}_{ob}(p_{yb}, y_b)$ and specify that $p^{\mathrm{I}}_{yo}=y^{\mathrm{I}}_o=0$, then

$$p^{\mathrm{I}}_{yb}=\frac{\partial U^{\mathrm{I}}_{ob}}{\partial y_b}, \quad y^{\mathrm{I}}_b=-\frac{\partial U^{\mathrm{I}}_{ob}}{\partial p_{yb}}. \tag{3.5.36}$$

Hence we find that if

$$U^{\mathrm{I}}_{ob}=\tfrac{1}{2}Lp^2_{yb}+Mp_{yb}y_b+\tfrac{1}{2}Ny^2_b, \tag{3.5.37}$$

then

$$\begin{pmatrix} A^{\mathrm{I}} & B^{\mathrm{I}} \\ C^{\mathrm{I}} & D^{\mathrm{I}} \end{pmatrix}=\begin{pmatrix} M & N \\ -L & -M \end{pmatrix}\begin{pmatrix} 1 & 0 \\ \mathsf{p}_0^{-1}(z_b-z_o) & 1 \end{pmatrix}. \tag{3.5.38}$$

Since the unperturbed ray may be written as

$$y(z)=\mathsf{p}_0^{-1}(z-z_b)p_{yb}+y_b, \tag{3.5.39}$$

it is not difficult to prove that

$$\left.\begin{aligned} L&=\ \mathsf{p}_0^{-2}(I-z_b^2I_0-2z_bI_1-I_2),\\ M&=\ \mathsf{p}_0^{-1}(z_bI_0-I_1),\\ N&=-I_0,\end{aligned}\right\} \tag{3.5.40}$$

where

$$I=\int_{z_o}^{z_b}(\mathsf{p}-\mathsf{p}_0)\,dz, \quad I_0=\int_{z_o}^{z_b}\mathsf{Y}\,dz, \quad I_1=\int_{z_o}^{z_b}\mathsf{Y}z\,dz, \quad I_2=\int_{z_o}^{z_b}\mathsf{Y}z^2\,dz. \tag{3.5.41}$$

Hence we may establish the formulae

$$\left.\begin{aligned} A^{\mathrm{I}} &= \mathsf{p}_0^{-1}(z_o I_0 - I_1), \\ B^{\mathrm{I}} &= -I_0, \\ C^{\mathrm{I}} &= -\mathsf{p}_0^{-2}(I - z_o z_b I_0 + (z_o + z_b) I_1 - I_2), \\ D^{\mathrm{I}} &= -\mathsf{p}_0^{-1}(z_b I_0 - I_1). \end{aligned}\right\} \tag{3.5.42}$$

It follows at once from (3.4.25) and (3.5.42) that

$$\frac{\mathrm{I}}{f_o} = \frac{\mathrm{I}}{f_b} = \mathsf{p}_0^{-1}\int_{z_o}^{z_b} \mathsf{Y}\, dz, \tag{3.5.43}$$

particular cases of which are (1.4.31) and (1.5.14), and from (3.4.27) and (3.5.42) that

$$z_{ho} = z_{hb} = \left(\int_{z_o}^{z_b} \mathsf{Y} z\, dz\right) \Big/ \left(\int_{z_o}^{z_b} \mathsf{Y}\, dz\right). \tag{3.5.44}$$

It is sometimes stated that, to this approximation, the principal planes coincide at the 'centre of gravity' of the field.

We are now in a position to specify the conditions which must be satisfied by the field in order that rays extending in the direction $z_b \to \infty$ should possess asymptotes. It is certainly necessary that p tends to a limit p_∞, say, and that $\mathsf{Y} \to 0$. Hence, by taking z_o sufficiently large, we may apply the weak-field approximation to the asymptotic region. If we now specify that $\mathsf{p}_0 = \mathsf{p}_\infty$ and $p_{yb}^{\mathrm{I}} = y_b^{\mathrm{I}} = 0$, where z_b is taken at infinity, the (fictitious) unperturbed ray is the asymptote of the (real) perturbed ray. It is clear from (3.4.24) that if $p_{yb} = 0$, so that the ray is asymptotically parallel to the axis, the co-ordinates (p_{yo}, y_o) of the asymptote are convergent if A and B are convergent, i.e., from (3.5.42), if the integrals I_0 and I_1 converge as $z_b \to \infty$. If the ray is not asymptotically parallel to the axis, it is also necessary that C and D should be convergent so that I and I_2 must then converge as $z_b \to \infty$.‡

It is possible to derive integral formulae, similar to (3.5.43) and (3.5.44) but more complicated, which give better approximations for the positions of the cardinal points. Svartholm§ has given such formulae, applicable to magnetic lenses, involving only the moments introduced in (3.5.41). A technique which is much favoured is that of solving the ray equation by the Picard iterative procedure,||

‡ For further discussion of this point, see ref. (24).

§ N. Svartholm, *Ark. Mat. Astr. Fys.* **35** A (1948), no. 6.

|| See, for instance, J. Picht, *Ann. Phys., Lpz.*, **15** (1932), 926–64.

but it is hard to see why the evaluation of a series of indefinite integrals should be preferred to the solution of a linear differential equation of the form (3.3.28). An exception is the application of the Picard technique to purely magnetic lenses, for one then obtains the solution of the ray equation expressed as a power series involving a parameter which represents the strength of the lens;‡ this may provide a convenient method for calculating the lens characteristics for a range of values of the energizing current.

‡ M. v. Ments and J. B. Le Poole, *Appl. Sci. Rec.* 1 (1950), 3–17.

CHAPTER 4

THE ROTATIONALLY SYMMETRICAL SYSTEM

4.1. Introduction

Now that we are better equipped for the analysis of the optical properties of electron-optical instruments, it is proposed that we undertake a more extensive investigation of the general rotationally symmetrical system than was possible in Chapter 1. Our calculations will, except in certain instances, be relativistically accurate but ignore the effect of space charge. An appropriate notation will make it possible to indicate the non-relativistic forms of important formulae.

The rules set out in § 3.3 will enable us to pass quickly over the paraxial properties. In discussing the third-order aberrations, however, it will be necessary to establish formulae for the coefficients which characterize these aberrations and also the optical significance of these coefficients.

The next step will be to investigate the chromatic aberration, that is, the optical effect of fluctuations in the beam energy. It would be possible to evaluate in just the same way the effects of fluctuations in the magnetic and electric fields, but these calculations will not be pursued. However, we shall show how the 'relativistic correction', that is, the difference between the relativistic and non-relativistic treatments, may be evaluated. Although the non-relativistic approximation is satisfactory only for beam energies below about 1000 V. and the relativistic treatment is essential for energies above about 100,000 V., it is possible between these limits to evaluate the relativistic correction by perturbation theory.

As will be seen in § 4.5, there is no difficulty in extending our calculations to take account of the existence of a *known* space-charge and space-current distribution. This extension is of interest, since it is possible to exert some control over certain aberrations by the use of 'controlled' space charge. It is also possible to calculate the effect upon an electron beam of its intrinsic space charge if the beam is assumed uniform in cross-section and if the calculations are carried through to the paraxial approximation only.

Since rotational symmetry itself guarantees Gaussian imaging in any electron-optical system, any slight departure from rotational symmetry will introduce aberrations. Of the various possible types of asymmetry, *ellipticity* is the most important. It proves possible to establish by elementary considerations formulae for the astigmatism which arises from ellipticity in electric and magnetic lenses. The study of asymmetries is of some importance to the designers of electron microscopes, since the resolving power is at present limited by errors in the machining of the objective lenses.

4.2. The paraxial imaging properties

It is proposed that, since the system is rotationally symmetrical, we adopt complex co-ordinates $(\bar{u}, u, z)$ at the outset. We find from (1.4.26) that the field equation (1.2.8) becomes

$$\nabla^2\phi \equiv \left\{\frac{\partial^2}{\partial z^2} + 4\frac{\partial^2}{\partial \bar{u}\,\partial u}\right\}\phi = 0 \tag{4.2.1}$$

in the absence of space charge. The general rotationally symmetrical solution of (4.2.1) may be expanded as

$$\phi = \Phi - \tfrac{1}{4}\bar{u}u\Phi'' + \tfrac{1}{64}\bar{u}^2u^2\Phi^{\text{iv}} - \dots, \tag{4.2.2}$$

in agreement with (1.4.3).

Although it would be possible to establish in complex co-ordinates a formula for the magnetic vector potential equivalent to (1.5.2), it will be convenient to introduce at this point a *complex potential* which is easier to handle than the three components of the vector potential and is simply related to both the vector potential and the scalar potential. However, the advantages of this procedure will not come to light until §§ 4.5 and 4.6, in which we shall consider space currents and asymmetrical fields.

The field equation (1.2.9) may be rewritten as

$$\text{grad div}\,\mathbf{A} - \nabla^2\mathbf{A} = 0, \tag{4.2.3}$$

if we assume that there is no space current. In addition to (4.2.3), it is usual to impose the condition

$$\text{div}\,\mathbf{A} = 0, \tag{4.2.4}$$

in virtue of which (4.2.3) reduces to

$$\nabla^2 \mathbf{A} = 0. \tag{4.2.5}$$

Written in complex co-ordinates, (4.2.4) becomes

$$\frac{\partial}{\partial \bar{u}}(A_x - iA_y) + \frac{\partial}{\partial u}(A_x + iA_y) + \frac{\partial A_z}{\partial z} = 0, \tag{4.2.6}$$

which is satisfied if $\mathbf{A}$ is derivable from a complex function of position $U(\bar{u}, u, z)$ according to the equations

$$A_x + iA_y = -\frac{\partial U}{\partial z}, \quad A_z = \frac{\partial \bar{U}}{\partial \bar{u}} + \frac{\partial U}{\partial u}. \tag{4.2.7}$$

Moreover, the field equation (4.2.3) is satisfied if the 'complex potential' U satisfies the Laplace equation

$$\nabla^2 U = 0. \tag{4.2.8}$$

It is not difficult to verify that ψ, defined by

$$\psi = i\left\{\frac{\partial \bar{U}}{\partial \bar{u}} - \frac{\partial U}{\partial u}\right\}, \tag{4.2.9}$$

is real, satisfies the Laplace equation, and leads to the same field components as $\mathbf{A}$ which satisfies (4.2.7) if $\psi(\bar{u}, u, z)$ is identified with the magnetic scalar potential.

It is obvious from (4.2.9) that U represents a field of rotational symmetry only if it transforms, on rotation of the co-ordinate system about the z-axis, in the same way as u. We therefore adopt the following solution of (4.2.8):

$$U = i\{\tfrac{1}{2}u\Psi - \tfrac{1}{16}\bar{u}u^2\Psi'' + \tfrac{1}{192}\bar{u}^2u^3\Psi^{\text{iv}} - \ldots\}, \tag{4.2.10}$$

in which $\Psi(z)$ may, in view of (4.2.9), be interpreted as the value of the magnetic scalar potential upon the z-axis.

It is not difficult to see from (1.3.10), (3.2.2) and (4.2.7) that the variational function is given by the formula

$$m = p\sqrt{(1 + \bar{u}'u')} + \frac{1}{2}\left\{\bar{u}'\frac{\partial U}{\partial z} + u'\frac{\partial \bar{U}}{\partial z}\right\} - \left\{\frac{\partial \bar{U}}{\partial \bar{u}} + \frac{\partial U}{\partial u}\right\}. \tag{4.2.11}$$

If this function is expanded, with the help of (1.2.4), (4.2.2) and (4.2.10), in the form (3.3.1) we find, as we might have expected, that all odd terms vanish, and that $m^{(2)}$ and $m^{(4)}$ are given by the formulae

$$m^{(2)} = \tfrac{1}{2}\mathsf{p}\bar{u}'u' - \tfrac{1}{4}\mathsf{p}^{-1}\langle 1+\Phi\rangle\Phi''\bar{u}u - \tfrac{1}{4}i\mathsf{H}(\bar{u}'u-\bar{u}u') \quad (4.2.12)$$

and

$$m^{(4)} = -\tfrac{1}{8}\mathsf{p}\bar{u}'^2u'^2 - \tfrac{1}{8}\mathsf{p}^{-1}\langle 1+\Phi\rangle\Phi''\bar{u}'u'\bar{u}u - \tfrac{1}{32}\mathsf{p}^{-3}\Phi''^2\bar{u}^2u^2$$
$$+\tfrac{1}{64}\mathsf{p}^{-1}\langle 1+\Phi\rangle\Phi^{\mathrm{iv}}\bar{u}^2u^2 + \tfrac{1}{32}i\mathsf{H}''(\bar{u}'u-\bar{u}u')\bar{u}u, \quad (4.2.13)$$

where $$\mathsf{p} = \sqrt{(2\Phi\langle 1+\tfrac{1}{2}\Phi\rangle)}, \quad (4.2.14)$$

and where, as in §1.5, $\mathsf{H}(z)$ is the magnetic field strength on the z-axis. We have used the relation $\mathsf{H} = -\Psi'$, and we have adopted the following rule in order to facilitate the reduction of relativistic formulae to their non-relativistic forms: *Terms enclosed by angular brackets are to be replaced by unity in the non-relativistic approximation.*

We may confirm by comparison of (3.3.24) and (4.2.12) that the latter is Gaussian so that we may proceed with the transformations (3.3.23). On writing

$$u = v\,e^{i\chi}, \quad (4.2.15)$$

where $$\frac{d\chi}{dz} = \tfrac{1}{2}\mathsf{p}^{-1}\mathsf{H}, \quad (4.2.16)$$

$m^{(2)}$ adopts the form (3.3.25), where

$$\mathsf{V} = \tfrac{1}{4}(2\mathsf{p}^{-1}\langle 1+\Phi\rangle\Phi'' + \mathsf{p}^{-1}\mathsf{H}^2). \quad (4.2.17)$$

The adoption of the further transformation

$$v = \mathsf{p}^{-\frac{1}{2}}w \quad (4.2.18)$$

will help to simplify the formulae which are needed for the numerical calculation of lens properties. Thus we find that W which appears in (3.3.26) is given by

$$\mathsf{W} = \tfrac{1}{4}(3\mathsf{p}^{-4}\langle 1+\tfrac{1}{3}\mathsf{p}^2\rangle\Phi'^2 + \mathsf{p}^{-2}\mathsf{H}^2), \quad (4.2.19)$$

and hence may be evaluated more easily and accurately than V which involves Φ''. The determination of the higher derivatives of an experimentally determined potential distribution is a most unsatisfactory operation which it is wise to avoid. The paraxial ray equation is now given by (3.3.28), i.e. by

$$\frac{d^2w}{dz^2} + \mathsf{W}w = 0. \quad (4.2.20)$$

If we now introduce object, aperture and image planes, we may write the general solution of (4.2.20) in the form

$$w(z) = w_o g(z) + w_a h(z), \tag{4.2.21}$$

where $g(z)$ and $h(z)$ are solutions of (4.2.20) satisfying the boundary conditions (3.3.32), i.e.

$$g_o = 1, \quad g_a = 0, \quad h_o = 0, \quad h_a = 1. \tag{4.2.22}$$

The imaging condition is (3.3.33), but (3.3.35) is replaced by

$$k = g'h - gh' \tag{4.2.23}$$

and (3.3.36) by $\quad k = -h'_o, \quad k = g'_a, \quad k = -g_b h'_b.$ (4.2.24)

Moreover, we must evaluate the magnification from

$$M = (\mathsf{p}_o/\mathsf{p}_b)^{\frac{1}{4}} g_b, \tag{4.2.25}$$

rather than (3.3.34), since this is the value for which

$$v_b = M v_o. \tag{4.2.26}$$

Apart from the advantage mentioned in the preceding paragraph and apart from the fact that (4.2.20) is easier to solve numerically than (3.3.27), the transformation (4.2.18) offers the further advantage that $g'(z)$ and $h'(z)$, which are always needed for aberration calculations, may easily be computed from the indefinite integration formulae

$$g'(z) = g'_o - \int_{z_o}^{z} \mathsf{W}(\zeta) g(\zeta)\, d\zeta, \quad h'(z) = h'_o - \int_{z_o}^{z} \mathsf{W}(\zeta) h(\zeta)\, d\zeta. \tag{4.2.27}$$

4.3. Third-order aberrations

We now turn to the problem of determining the influence upon the imaging properties of the higher terms $m^{(4)}$, $m^{(6)}$, etc., in the expansion of m. If these properties are to approximate to those of Gaussian dioptrics, the field-of-view and aperture must be chosen so that the higher terms are much smaller than $m^{(2)}$. However, $m^{(4)}$ will then be much larger than successive terms, so that a study of the corresponding 'third-order' aberrations will acquaint us with the most important limitations of an instrument operating at finite field-of-view and aperture.

Let us consider the fictitious perturbation specified by

$$\mathsf{P}m = m^{(2)} + \rho m^{(4)} + \rho^2 m^{(6)} + \dots. \tag{4.3.1}$$

The 'unperturbed' system is the Gaussian approximation to the real system which is represented by $\mathsf{P}m$ when $\rho = 1$. The parameter ρ separates aberrations according to their dependence upon the off-axis co-ordinates, and it is obvious that we may evaluate the third-order aberrations by calculating the first-order effect of the fictitious perturbation. If we introduce the appropriate aberration characteristic function

$$V_{ob}^{(4)} = \int_{z_o}^{z_b} m^{(4)}\,dz, \tag{4.3.2}$$

it follows from (3.5.25) that the third-order aberration may be derived from $V_{ob}^{(4)}(\bar{w}_o, w_o, \bar{w}_a, w_a)$ by means of the formula

$$w_b^{(3)} = 2k^{-1}g_b \frac{\partial V_{ob}^{(4)}}{\partial \bar{w}_a}. \tag{4.3.3}$$

It is proposed that we establish and discuss the general form of the third-order aberrations before we obtain explicit formulae for their coefficients.

Since $V_{ob}^{(4)}$ is a scalar function which characterizes the aberrations of a system of rotational symmetry, it must be unchanged when w is changed to $w\,e^{i\chi}$, corresponding to a rotation of the co-ordinate axes about the axis of symmetry. Hence $V_{ob}^{(4)}$ has the form‡

$$V_{ob}^{(4)} = \begin{pmatrix} \bar{w}_o^2 \\ \bar{w}_o\bar{w}_a \\ \bar{w}_a^2 \end{pmatrix}' \begin{pmatrix} K^* & \bar{R}^* & \bar{Q}^* \\ R^* & L^* & \bar{P}^* \\ Q^* & P^* & N^* \end{pmatrix} \begin{pmatrix} w_o^2 \\ w_o w_a \\ w_a^2 \end{pmatrix}, \tag{4.3.4}$$

where K^*, L^* and N^* are real but P^*, Q^* and R^* are possibly complex. If the field is purely electric, the co-ordinate system does not suffer the rotation (4.2.15); moreover, we see from (4.2.2) and (4.2.11) that the variational function is unaffected by an interchange of u and $\bar{u}$, i.e. by the adoption of a left-hand set of axes instead of a right-hand set. The same must be true of the characteristic functions, so that *if the field is purely electric, all coefficients of the third-order aberrations are real.*

The formulae (4.3.3) and (4.3.4) give immediately an expression for $w_b^{(3)}$. However, we shall wish to evaluate the aberrations not in the modified scale in which w is measured but in the original scale

‡ In matrix equations, a prime denotes the transpose of a matrix or vector.

of u and v. We should therefore reverse the transformation (4.2.18) to obtain the formula

$$v_b^{(3)} = \begin{pmatrix} \bar{v}_o \\ 2\bar{v}_a \end{pmatrix}' \begin{pmatrix} R & L & \bar{P} \\ Q & P & N \end{pmatrix} \begin{pmatrix} v_o^2 \\ v_o v_a \\ v_a^2 \end{pmatrix}, \qquad (4\cdot3\cdot5)$$

where

$$\left.\begin{aligned} L &= 2k^{-1}M\mathsf{p}_o^{\frac{1}{2}}\mathsf{p}_a^{\frac{1}{2}}L^*, \quad N = 2k^{-1}M\mathsf{p}_o^{-\frac{1}{2}}\mathsf{p}_a^{\frac{3}{2}}N^*, \\ P &= 2k^{-1}M\mathsf{p}_a P^*, \quad Q = 2k^{-1}M\mathsf{p}_o^{\frac{1}{2}}\mathsf{p}_a^{\frac{1}{2}}Q^*, \quad R = 2k^{-1}M\mathsf{p}_o R^*. \end{aligned}\right\} \qquad (4\cdot3\cdot6)$$

The aberration coefficients L, N, P, Q and R are the coefficients of field curvature, spherical aberration, coma, astigmatism and distortion, respectively. We may ascribe values to the coefficients which are 'referred back' to the object plane by omitting the magnification M from (4.3.6).

It is interesting to consider the dependence of these coefficients upon the aperture position in order to determine whether, by an appropriate choice of this position, any of the coefficients may be made to vanish. Let us therefore suppose that if the aperture is displaced from z_a to $z_{\tilde{a}}$, L, etc., are changed to $\tilde{L}$, etc. We may express $w_{\tilde{a}}$ in terms of w_o and w_a by (4.2.21), and hence, with the help of (4.2.18), relate v_a to v_o and $v_{\tilde{a}}$. If we now equate $v_b^{(3)}$ given by (4.3.5) to $v_b^{(3)}$ given by the 'displaced' form of (4.3.5), we obtain the relations

$$\left.\begin{aligned} \tilde{L} &= \mathsf{p}_{\tilde{a}}^{\frac{1}{2}}\mathsf{p}_a^{-\frac{1}{2}}h_{\tilde{a}}^{-1}L - 2\mathsf{p}_o^{\frac{1}{2}}\mathsf{p}_a^{-\frac{1}{2}}g_{\tilde{a}}h_{\tilde{a}}^{-1}(\bar{P}+P) + 4\mathsf{p}_o\mathsf{p}_{\tilde{a}}^{\frac{1}{2}}\mathsf{p}_a^{-\frac{3}{2}}h_{\tilde{a}}^{-3}N, \\ \tilde{N} &= \mathsf{p}_{\tilde{a}}^{\frac{3}{2}}\mathsf{p}_a^{-\frac{3}{2}}h_{\tilde{a}}^{-3}N, \\ \tilde{P} &= \mathsf{p}_{\tilde{a}}\mathsf{p}_a^{-1}h_{\tilde{a}}^{-2}P - 2\mathsf{p}_o^{\frac{1}{2}}\mathsf{p}_{\tilde{a}}\mathsf{p}_a^{-\frac{3}{2}}g_{\tilde{a}}h_{\tilde{a}}^{-3}N, \\ \tilde{Q} &= \mathsf{p}_{\tilde{a}}^{\frac{1}{2}}\mathsf{p}_a^{-\frac{1}{2}}h_{\tilde{a}}^{-1}Q - \mathsf{p}_o^{\frac{1}{2}}\mathsf{p}_{\tilde{a}}^{\frac{1}{2}}\mathsf{p}_a^{-1}g_{\tilde{a}}h_{\tilde{a}}^{-2}P + \mathsf{p}_o\mathsf{p}_{\tilde{a}}^{\frac{1}{2}}\mathsf{p}_a^{-\frac{3}{2}}g_{\tilde{a}}^2h_{\tilde{a}}^{-3}N, \\ \tilde{R} &= R - \mathsf{p}_o^{\frac{1}{2}}\mathsf{p}_a^{-\frac{1}{2}}g_{\tilde{a}}h_{\tilde{a}}^{-1}L + \mathsf{p}_o\mathsf{p}_a^{-1}g_{\tilde{a}}^2h_{\tilde{a}}^{-2}(\bar{P}+2P) \\ &\qquad - 2\mathsf{p}_o^{\frac{1}{2}}\mathsf{p}_a^{-\frac{1}{2}}g_{\tilde{a}}h_{\tilde{a}}^{-1}Q - 2\mathsf{p}_o^{\frac{3}{2}}\mathsf{p}_a^{-\frac{3}{2}}g_{\tilde{a}}^3h_{\tilde{a}}^{-3}N. \end{aligned}\right\} \qquad (4\cdot3\cdot7)$$

The one drawback to the adoption of the transformation (4.2.18) is evident in (4.3.7); the factors p would not have appeared had we not adopted (4.2.21) in preference to (3.3.31) as the equation of the general paraxial ray. It is therefore worth noticing that if the functions $g(z)$, $h(z)$ and the invariant k which appear in (3.3.31) and (3.3.35) are labelled by the suffix v, then

$$g_v = (\mathsf{p}/\mathsf{p}_o)^{-\frac{1}{2}}g, \quad h_v = (\mathsf{p}/\mathsf{p}_a)^{-\frac{1}{2}}h \quad \text{and} \quad k_v = \mathsf{p}_o^{\frac{1}{2}}\mathsf{p}_a^{\frac{1}{2}}k. \qquad (4\cdot3\cdot8)$$

Now if $z_o < z_a < z_b$, $g_{\bar{a}}/h_{\bar{a}}$ takes all values between $-\infty$ and ∞ as $z_{\bar{a}}$ varies between z_o and z_b. It follows from (4.3.7) that, if we write

$$P = P_r + iP_i, \quad \text{etc.,} \tag{4.3.9}$$

P_r and Q_i can definitely be made to vanish and that N and P_i can definitely not be made to vanish merely by adjusting the aperture position. Whether or not the other coefficients can be eliminated depends upon the relative magnitudes and signs of the coefficients.‡

We see from the second of equations (4.3.7) that the change in the spherical aberration coefficient due to the displacement of the aperture plane is purely formal in the sense that if the aperture of a beam is measured by a scale which is not directly related to the aperture plane—for instance, by the angle subtended by the beam at the object—then N is independent of the aperture position. The same is not true of the other coefficients, but it is true of a certain combination of the coefficients of field curvature and astigmatism, for we easily see from (4.3.7) that

$$\tilde{L} - 4\tilde{Q}_r = \mathsf{p}_{\bar{a}}^{\frac{1}{2}} \mathsf{p}_a^{-\frac{1}{2}} h_{\bar{a}}^{-1} (L - 4Q_r). \tag{4.3.10}$$

We also see that $L - 4Q_r$, which is a form of the *Petzval coefficient*, can definitely not be made to vanish by adjustment of the aperture position.

Let us now investigate the geometrical significance of the various aberration coefficients by determining their influence upon the image of a point object. Let the object be at $v_o = r_o$, and let an arbitrary point of the aperture plane be written as $v_a = r_a e^{i\theta_a}$. If the radius of the diaphragm is r_A, we need consider only rays for which $r_a \leqslant r_A$.

It is convenient to consider together the coefficient of *field curvature* L, which is real, and the coefficient of *astigmatism* Q, which may be complex if the field is not purely electric. Then, if N, P and R all vanish, the ray through the object point and an arbitrary point of the aperture intersects the Gaussian image plane at the point with complex co-ordinate $Mr_o + v_b^{(3)}$, where, from (4.3.5),

$$v_b^{(3)} = (L e^{i\theta_a} + 2Q e^{-i\theta_a}) r_o^2 r_a. \tag{4.3.11}$$

‡ For further discussion of this problem see H. Voit, *Z. Instrumkde*, **59** (1939), 71–82.

Let us ignore the term Mr_o which represents pure Gaussian image formation and consider the figure which is mapped in the image plane as θ_a varies between 0 and 2π and r_a varies between 0 and r_A. It is obvious that we shall obtain the envelope of this figure if we replace r_a in (4.3.11) by r_A. If we now agree to write

$$Q = |Q| e^{i\theta_q}, \quad \text{etc.,} \tag{4.3.12}$$

for the complex coefficients, the envelope is given by

$$v_b^{(3)} = \{L e^{i(\theta_a - \frac{1}{2}\theta_q)} + 2|Q| e^{-i(\theta_a - \frac{1}{2}\theta_q)}\} r_o^2 r_A e^{i\frac{1}{2}\theta_q}. \tag{4.3.13}$$

This figure is an ellipse with semi-axis $|L+2|Q||$ in the direction inclined at an angle $\frac{1}{2}\theta_q$ to the real axis, i.e. to the line joining the Gaussian image point to the centre of the image plane, and whose other semi-axis is $|L-2|Q||$.

We should now proceed to determine what image is obtained in planes slightly removed from the image plane in order to see whether a smaller image than the above could be obtained by refocusing or by adopting some image surface other than a plane. Let us therefore consider the intersection of the general ray with the plane $z = z_b + \delta z_b$; there will be a slight change in the paraxial magnification, but this is of no consequence. Since $h(z)$ is not zero in the new plane, we should add to the right-hand side of (4.3.13)

$$\mathsf{p}_b^{-\frac{1}{2}} h_b' \delta z_b \mathsf{p}_a^{\frac{1}{2}} v_a,$$

which, in view of (4.2.24) and (4.2.25), may be rewritten as

$$-\mathsf{p}_o^{\frac{1}{2}} \mathsf{p}_a^{\frac{1}{2}} \mathsf{p}_b^{-1} M^{-1} k \delta z_b r_A e^{i\theta_a}.$$

This is equivalent to replacing L by $L - Mk\mathsf{p}_o^{\frac{1}{2}} \mathsf{p}_a^{\frac{1}{2}} \mathsf{p}_b^{-1} \delta z_b r_b^{-2}$, if $r_b = Mr_o$.

Since by varying z_b we are able to vary the effective value of L, we are able to vary the dimensions of the ellipse which is the image of the object point. It is obvious that the best focus is obtained when L is effectively zero, for then the ellipse becomes a circle of radius $2|Q| r_o^2 r_A$. In this way we obtain the equation of the *surface of least confusion*.

$$z = z_b + M^{-1} k^{-1} \mathsf{p}_o^{-\frac{1}{2}} \mathsf{p}_a^{-\frac{1}{2}} \mathsf{p}_b L r_b^2. \tag{4.3.14}$$

If we replace L in (4.3.14) by $L-2|Q|$ or $L+2|Q|$, we obtain the surfaces in which the ellipse degenerates into a straight line of length $8|Q| r_o^2 r_A$ inclined at an angle $\frac{1}{2}\theta_q$ or $\frac{1}{2}(\pi - \theta_q)$, respectively, to the real axis. The separation between these surfaces at a given radius r_b, $4\mathsf{p}_o^{-\frac{1}{2}}\mathsf{p}_a^{-\frac{1}{2}}\mathsf{p}_b\,|M^{-1}k^{-1}Q|\,r_b^2$, is known as the *astigmatic difference*. If Q is real, the directions of the degenerate ellipses are radial—or 'sagittal'—and tangential; the associated surfaces are known as the 'sagittal' and 'tangential' surfaces. By replacing L

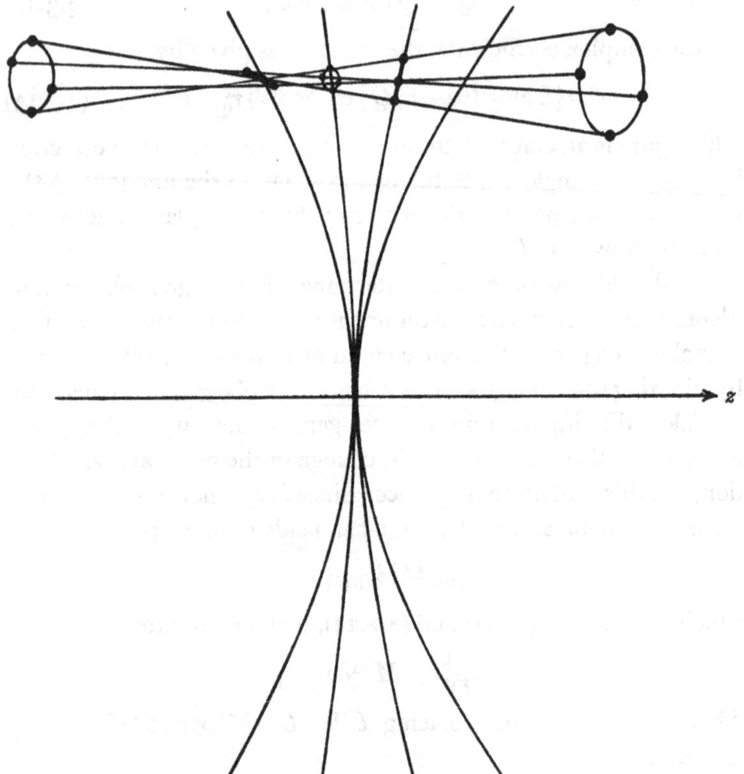

Fig. 4.1. Astigmatism and field curvature.

in (4.3.14) by the Petzval coefficient $L-4Q_r$ we obtain a fourth surface, known as the *Petzval surface*, which may be located as soon as the Petzval coefficient has been determined. If Q is real, the surfaces are ordered as follows: tangential surface, surface of least confusion, sagittal surface and Petzval surface; moreover, the curvatures of these surfaces are in arithmetic progression (fig. 4.1).

If we now suppose that only the *spherical aberration* coefficient N is non-zero, (4.3.5) reduces to

$$v_b^{(3)} = 2Nr_a^3\,e^{i\theta_a}, \tag{4.3.15}$$

from which it is clear that the geometrical pattern characteristic of this coefficient is independent of the object position and is rotationally symmetrical, i.e. circular. Any object point will have as its image in the Gaussian image plane a disk of radius $2\,|N|\,r_A^3$.

Let us now consider, as before, the form taken by the pattern in planes neighbouring the Gaussian image plane in order to determine whether the spherical-aberration disk can be reduced in size. In the plane $z = z_b + \delta z_b$, we obtain

$$v_b^{(3)} = (2Nr_a^3 - \mathrm{p}_o^{\frac{1}{2}}\mathrm{p}_a^{\frac{1}{2}}\mathrm{p}_b^{-1}kM^{-1}\delta z_b r_a)\,e^{i\theta_a}, \tag{4.3.16}$$

if we include the 'defocusing' term represented by h in (4.2.21) but ignore the change of magnification which we obtain from the term g. Since both terms on the right-hand side of (4.3.16) depend in the same way upon θ_a, we may ignore this variable and consider only $r_b^{(3)}$.

The problem is now to determine for what value of δz_b the maximum value of $r_b^{(3)}$, for the range $0 \leqslant r_a \leqslant r_A$, is a minimum. By introducing fresh variables, the problem could be identified with that of determining for what value of λ the maximum value of $|x^3 - \lambda x|$, in the range $0 \leqslant x \leqslant 1$, is a minimum. It is not difficult to verify that the required value of λ is $\frac{3}{4}$, i.e. three-quarters of the value which makes $|x^3 - \lambda x| = 0$ when x takes its maximum value, and that $|x^3 - \lambda x|$ then has a maximum of $\frac{1}{4}$, i.e. one-quarter of its maximum value when $\lambda = 0$. We now see at once that the radius of the spherical-aberration disk will be reduced by a factor of four to $\frac{1}{2}\,|N|\,r_A^3$ when the image plane is displaced three-quarters of the way towards the 'marginal' focal plane in which $r_b^{(3)} = 0$ when $r_a = r_A$, i.e. by an amount

$$\delta z_b = \tfrac{3}{2}k^{-1}\mathrm{p}_o^{-\frac{1}{2}}\mathrm{p}_a^{-\frac{1}{2}}\mathrm{p}_b MNr_A^2. \tag{4.3.17}$$

Fig. 4.2 will help to clarify the relationship between the Gaussian focus, the marginal focus and the best focus. It may again be noted that it is our adoption of (4.2.18) which has led to the appearance of the factor $\mathrm{p}_o^{\frac{1}{2}}\mathrm{p}_a^{\frac{1}{2}}\mathrm{p}_b^{-1}$ multiplying k in (4.3.14) and (4.3.17).

Since, as has already been observed, the spherical aberration is, in a sense, independent of the aperture position, it is convenient to express the minimum radius of the spherical-aberration disk in the form

$$r_b^{(3)} = \tfrac{1}{4} \, | MC_s | \, \omega_o^3, \tag{4.3.18}$$

where ω_o is the semi-aperture angle of rays at the object. The new coefficient C_s is related to N^* by

$$C_s = -4\mathsf{p}_o k^{-4} N^*. \tag{4.3.19}$$

It is not difficult to verify from the above paragraph that the marginal focus is displaced along the z-axis from the Gaussian

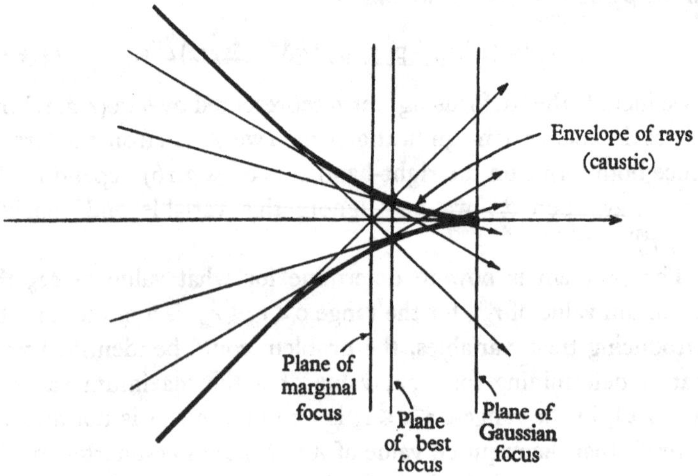

Fig. 4.2. Relationship between the various focal planes.

focus by a distance $-M^2(\mathsf{p}_b/\mathsf{p}_o)\, C_s \omega_o^2$, so that if we say that the spherical aberration is 'positive' when C_s is positive, the spherical aberration is positive or negative according as the marginal focus does or does not lie on the object side of the Gaussian focus. We see from (4.3.18) that the minimum radius of the spherical-aberration disk, referred back to the object plane, is $\tfrac{1}{4} \, | C_s | \, \omega_o^3$, and we also find, on recalling the formula (2.5.26) or (3.4.9) for the longitudinal magnification, that the displacement of the marginal from the Gaussian focus—the 'longitudinal spherical aberration'—has the value $-C_s \omega_o^2$ when referred back to the object plane.

We shall now proceed to find the image pattern associated with the coefficient of *coma* P by assuming that all other coefficients vanish. Then (4.3.5) reduces to

$$v_b^{(3)} = 2Pv_o\bar{v}_a v_a + \bar{P}\bar{v}_o v_a^2 \tag{4.3.20}$$

which, with the help of the notation of (4.3.12), leads to

$$v_b^{(3)} = |P|\,(2 + e^{2i(\theta_a - \theta_p)})\, r_o r_a^2 e^{i\theta_p}. \tag{4.3.21}$$

As θ_a varies from 0 to 2π, the term of (4.3.21) enclosed in brackets traces and retraces a circle of unit radius whose centre is situated on the real axis two units from the origin. If we now allow for the other terms of (4.3.21) and for the fact that r_a varies between 0 and r_A, we see that the complete coma pattern resembles a comet

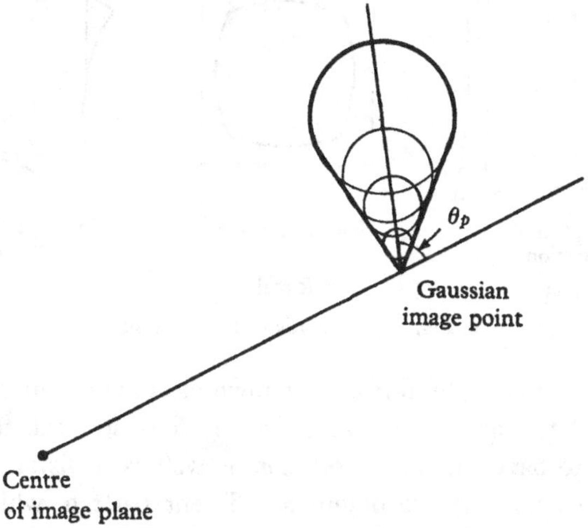

Fig. 4.3. The coma pattern.

or a dart whose tail will be situated at the Gaussian image point and whose head is inclined at an angle θ_p or $\pi + \theta_p$ to the radius vector from the centre of the image plane to the image point according as the magnification M is positive or negative (fig. 4.3). The overall length of the coma pattern is $3\,|P|\,r_o r_A^2$.

If we now suppose that all coefficients except the *distortion* coefficient R vanish, (4.3.5) reduces to

$$v_b^{(3)} = R\bar{v}_o v_o^2, \tag{4.3.22}$$

which, being independent of v_a, represents merely a *displacement* of the image point from its Gaussian position. We should therefore investigate not the image of a point object but the image of an extended object such as, for instance, a square. If we write $v_o = r_o\,e^{i\theta_o}$, the co-ordinate of the image point is given by

$$v_b = Mr_o\,e^{i\theta_o} + |R|\,r_o^3\,e^{i(\theta_o+\theta_r)}. \tag{4.3.23}$$

By determining the locus of this point when v_o traces a square whose centre is at the origin, we find that if the argument (or 'phase') of MR lies between $-\frac{1}{2}\pi$ and $\frac{1}{2}\pi$ the square becomes concave, and if the argument lies between $\frac{1}{2}\pi$ and $\frac{3}{2}\pi$ the square

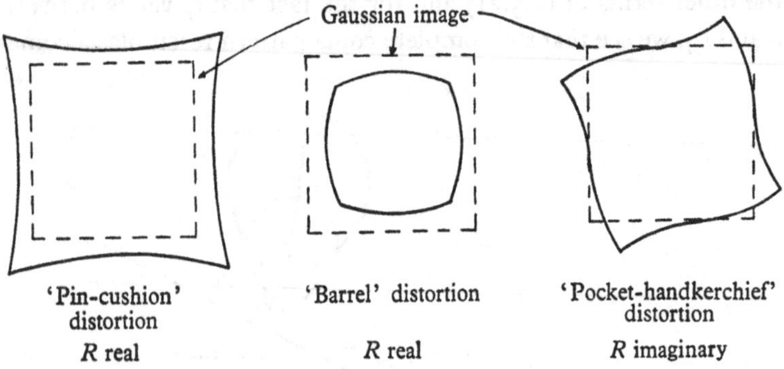

Fig. 4.4. The distortion patterns.

becomes convex; the former is known as 'pincushion distortion' and the latter as 'barrel distortion'. If R is not real, the square appears to have suffered a rotation as well as a distortion. If R is purely imaginary we obtain a different pattern, which one is tempted to name 'pocket-handkerchief distortion'. (See fig. 4.4.)

It now remains for us to establish formulae for the coefficients of third-order aberration by evaluating the integral (4.3.2). The first step will be to transform the function $m^{(4)}(\bar{u}', u', \bar{u}, u, z)$ given by (4.2.13) into a function of the variables $\bar{w}'$, w', $\bar{w}$ and w by means of (4.2.15), (4.2.16) and (4.2.18). We may, however, go further for it proves possible, by repeated partial integrations, to reduce the function to the form

$$m^{(4)} = \frac{d}{dz}\{\mathsf{p}^{-1}(\mathscr{A}\Theta\Upsilon' + \mathscr{B}\Theta\Upsilon + \mathscr{C}\Upsilon\Upsilon' + \mathscr{D}\Upsilon'\Omega + \mathscr{E}\Upsilon\Omega - \tfrac{1}{2}\mathscr{B}\Omega^2)\}$$
$$+\mathsf{p}^{-1}(A\Theta\Upsilon + B\Upsilon^2 + C\Theta\Omega + D\Upsilon\Omega + E\Omega^2), \tag{4.3.24}$$

where $\quad \Theta = \bar{w}'w', \quad \Upsilon = \bar{w}w \quad$ and $\quad \Omega = i(\bar{w}'w - \bar{w}w'), \qquad (4.3.25)$

and $\mathscr{A}, \ldots, E$ are real functions of z. The factor p^{-1} has been included for convenience and the integrated term in Υ^2 has been ignored since, by the partial-integration rule, it is of no significance; the recurrence of the coefficient $\mathscr{B}$ may be regarded as fortuitous. The aim in effecting the reduction (4.3.24) has been to reduce to five the number of functions—$A, \ldots, E$—which need be evaluated between the limits of integration (the coefficients $\mathscr{A}, \ldots, \mathscr{E}$ need be known only at the limits of integration) and to eliminate all derivatives of Φ and H higher than Φ'' and H'.

The reader may be spared details of the tedious and uninteresting reduction and acquainted only with the final formulae for the coefficients appearing in (4.3.24). These are:

$$\left.\begin{aligned}
\mathscr{A} &= -\tfrac{1}{16},\\
\mathscr{B} &= \tfrac{1}{16}\mathsf{p}^{-2}\langle 1+\Phi\rangle\Phi',\\
\mathscr{C} &= -\tfrac{1}{32}\{\mathsf{p}^{-4}\langle 1+\mathsf{p}^2\rangle\Phi'^2+\mathsf{p}^{-2}\langle 1+\Phi\rangle\Phi''+\mathsf{p}^{-2}\mathsf{H}^2\},\\
\mathscr{D} &= -\tfrac{1}{32}\mathsf{p}^{-1}\mathsf{H},\\
\mathscr{E} &= \tfrac{1}{32}\mathsf{p}^{-1}\mathsf{H}',
\end{aligned}\right\} \qquad (4.3.26)$$

and

$$\left.\begin{aligned}
A &= \tfrac{1}{32}\{3\mathsf{p}^{-4}\langle 1+\tfrac{1}{3}\mathsf{p}^2\rangle\Phi'^2+\mathsf{p}^{-2}\mathsf{H}^2\},\\
B &= \tfrac{1}{128}\{-118\mathsf{p}^{-8}\langle 1+\tfrac{117}{118}\mathsf{p}^2+\tfrac{15}{118}\mathsf{p}^4\rangle\Phi'^4\\
&\quad +65\mathsf{p}^{-6}\langle 1+\tfrac{3}{13}\mathsf{p}^2\rangle\langle 1+\Phi\rangle\Phi'^2\Phi''\\
&\quad -10\mathsf{p}^{-4}\langle 1+\tfrac{2}{5}\mathsf{p}^2\rangle\Phi''^2\\
&\quad -27\mathsf{p}^{-6}\langle 1+\tfrac{2}{3}\mathsf{p}^2\rangle\Phi'^2\mathsf{H}^2-3\mathsf{p}^{-4}\langle 1+\Phi\rangle\Phi''\mathsf{H}^2\\
&\quad +18\mathsf{p}^{-4}\langle 1+\Phi\rangle\Phi'\mathsf{H}\mathsf{H}'-3\mathsf{p}^{-4}\mathsf{H}^4-4\mathsf{p}^{-2}\mathsf{H}'^2\},\\
C &= -\tfrac{1}{16}\mathsf{p}^{-1}\mathsf{H},\\
D &= \tfrac{1}{64}\{-5\mathsf{p}^{-5}\langle 1+\tfrac{3}{5}\mathsf{p}^2\rangle\Phi'^2\mathsf{H}-4\mathsf{p}^{-3}\langle 1+\Phi\rangle\Phi''\mathsf{H}\\
&\quad +4\mathsf{p}^{-3}\langle 1+\Phi\rangle\Phi'\mathsf{H}'-3\mathsf{p}^{-3}\mathsf{H}^3\},\\
E &= -\tfrac{1}{64}\{5\mathsf{p}^{-4}\langle 1+\tfrac{3}{5}\mathsf{p}^2\rangle\Phi'^2+3\mathsf{p}^{-2}\mathsf{H}^2\}.
\end{aligned}\right\} \qquad (4.3.27)$$

It may be noted that in the non-relativistic approximation, when all terms in angular brackets may be ignored, each term of the above formulae is made up of products of $\mathsf{p}^{-2}\Phi'$, $\mathsf{p}^{-1}\mathsf{H}$, etc.; and also that $\mathscr{D}$, $\mathscr{E}$, C and D vanish when the field is purely electric, as might have been predicted from an examination of (4.3.24). Moreover, from a comparison of (4.2.19) and (4.3.27), we see that $A=\tfrac{1}{8}\mathsf{W}$.

The second step in the evaluation of the coefficients is to use (4.2.21) to express Θ, Υ and Ω in terms of $\overline{w}_o$, $\overline{w}_a$, w_o, w_a and so to express $V^{(4)}_{ob}$, given by (4.3.2), in the form (4.3.4). In this way we obtain the formulae

$$\left.\begin{aligned}
L^* &= -4k[\mathrm{p}^{-1}\mathscr{A}g'h']_o^b + 4\int_{z_o}^{z_b} \mathrm{p}^{-1}gh(Ag'h'+Bgh)\,\mathrm{d}z \\
&\quad + k^2\int_{z_o}^{z_b} \mathrm{p}^{-1}(A+2E)\,\mathrm{d}z, \\
N^* &= \int_{z_o}^{z_b} \mathrm{p}^{-1}h^2(Ah'^2+Bh^2)\,\mathrm{d}z, \\
P^* &= -k[\mathrm{p}^{-1}\mathscr{A}h'^2]_o^b + 2\int_{z_o}^{z_b} \mathrm{p}^{-1}gh(Ah'^2+Bh^2)\,\mathrm{d}z \\
&\quad + k\int_{z_o}^{z_b} \mathrm{p}^{-1}Ahh'\,\mathrm{d}z - \mathrm{i}k\int_{z_o}^{z_b} \mathrm{p}^{-1}(Ch'^2+Dh^2)\,\mathrm{d}z, \\
Q^* &= k[\mathrm{p}^{-1}(-\mathscr{A}g'h' + \tfrac{1}{2}\mathscr{B}k + \mathrm{i}\mathscr{D}k)]_o^b + \int_{z_o}^{z_b} \mathrm{p}^{-1}gh(Ag'h'+Bgh)\,\mathrm{d}z \\
&\quad - k^2\int_{z_o}^{z_b} \mathrm{p}^{-1}E\,\mathrm{d}z - \mathrm{i}k\int_{z_o}^{z_b} \mathrm{p}^{-1}(Cg'h'+Dgh)\,\mathrm{d}z, \\
R^* &= -k[\mathrm{p}^{-1}(3\mathscr{A}g'^2 + \mathscr{B}gg' + \mathscr{C}g^2 + 2\mathrm{i}\mathscr{D}gg' + \mathrm{i}\mathscr{E}g^2)]_o^b \\
&\quad + 2\int_{z_o}^{z_b} \mathrm{p}^{-1}g^2(Ag'h'+Bgh)\,\mathrm{d}z + k\int_{z_o}^{z_b} \mathrm{p}^{-1}Agg'\,\mathrm{d}z \\
&\quad - \mathrm{i}k\int_{z_o}^{z_b} \mathrm{p}^{-1}(Cg'^2+Dg^2)\,\mathrm{d}z.
\end{aligned}\right\} \tag{4.3.28}$$

We see at once that if the field is purely electric the above coefficients are all real. We may also note that if the object and image planes are in field-free space only the coefficient $\mathscr{A}$ of (4.3.26) is non-zero, and that if the field is purely electric and the treatment non-relativistic, or if the field is purely magnetic, the formulae (4.3.27) simplify considerably.

It is interesting to reconsider the Petzval coefficient $L-4Q_r$, for which we may establish the formula

$$\begin{aligned}
L^* - 4Q_r^* &= -\tfrac{1}{8}k^2[\mathrm{p}^{-3}\langle 1+\Phi\rangle\Phi']_o^b \\
&\quad - \tfrac{1}{8}k^2\int_{z_o}^{z_b} \mathrm{p}^{-1}\{3\mathrm{p}^{-4}\langle 1+\tfrac{2}{3}\mathrm{p}^2\rangle\Phi'^2 + 2\mathrm{p}^{-2}\mathrm{H}^2\}\,\mathrm{d}z.
\end{aligned} \tag{4.3.29}$$

It is noteworthy that the above formula does not involve the higher derivatives Φ'' and H' nor the paraxial rays $g(z)$ and $h(z)$. It is clear, moreover, that *if the electric field strength vanishes at the object and image planes, the Petzval coefficient cannot vanish.* If the coefficient is to vanish, we must have $\Phi_b' < \Phi_o'$.

Let us finally return to the spherical-aberration coefficient. We find from (4.3.6), (4.3.19) and (4.3.28) that

$$C_s = -\mathsf{p}_o \int_{z_o}^{z_b} \mathsf{p}^{-1} \bar{h}^2 (A\bar{h}'^2 + B\bar{h}^2)\, dz, \tag{4.3.30}$$

where $\bar{h}(z) = -k^{-1}h(z)$, i.e. $\bar{h}(z)$ is the solution of (4.2.20) satisfying the boundary conditions

$$\bar{h}_o = 0, \quad \bar{h}_o' = 1. \tag{4.3.31}$$

We may establish a *thin-lens approximation* for C_s by assuming that the aperture $z = z_a$ is situated at the lens so that $\bar{h}$ may be replaced by $z_a - z_o$ in the integral (4.3.30), the derivative $\bar{h}'$ being neglected, and by taking into account only terms of the lowest order in the field strengths. In this way we obtain the non-relativistic formula

$$C_s = \tfrac{1}{64}(z_a - z_o)^4 \left\{ 5\mathsf{P}^{-4} \int_{-\infty}^{\infty} \Phi''^2 dz + 2\mathsf{p}^{-2} \int_{-\infty}^{\infty} \mathsf{H}'^2 dz \right\}. \tag{4.3.32}$$

One may verify from a comparison of (4.3.32) with (1.4.31) and (1.5.14) that if two thin lenses, one electric and one magnetic, have the same focal length and the same axial field distribution, the spherical aberration coefficient is slightly smaller for the electric lens than for the magnetic lens. However, an investigation of the terms appearing in the formula for B in (4.3.27) indicates that if the lenses are not weak one must expect the spherical aberration to be more severe in the electric lens.

It was shown by Scherzer‡ in 1936, by a series of partial integrations, that, to the non-relativistic approximation, the integral appearing in (4.3.30) is essentially negative and hence that *the spherical aberration is always positive in a lens which is (a) static, (b) rotationally symmetrical and (c) free of space charges, etc., in the neighbourhood of the axis.* There is no reason to believe that the theorem does not hold also when the treatment is relativistic, but Scherzer§ has more recently demonstrated that if any one of the above three conditions is relaxed both the spherical aberration

‡ Ref. (27). § Ref. (28).

and the chromatic aberration—which is to be considered in the next section—may be eliminated.

Of all the third-order aberrations, spherical aberration is much the most important for—in conjunction with diffraction effects‡—it is responsible for the limit set upon the resolving power of electron microscopes. In theory the coefficient can be reduced indefinitely by suitably shaping the field; since its dimensions are the inverse of those of length, the coefficient can be reduced by confining the field to as small a region as possible. However, the field must then become more and more intense so that the minimum value of the coefficient attainable in practice is determined by the maximum field strengths which can be achieved.§

Extensive lists of publications upon the third-order aberrations may be found in the text-books. The aberrations have been studied experimentally, computationally,|| and by the use of models which may be treated analytically.‡

4.4. Chromatic aberration and the relativistic correction

The calculations of the preceding two sections were based on the assumption that the beam is monochromatic, but it is obvious that this condition can be only approximately realized. We shall therefore proceed to derive formulae for coefficients which characterize the change in the image-forming properties of the system due to a small variation in the beam energy. Since—as was pointed out in § 1.2—electrons of various energies may be considered independently, we may in this way establish the characteristics of an image formed by a beam of electrons whose energies cover a narrow band.

Let us suppose that the energy of electrons is increased from ϕ to $\phi+\epsilon$; then we may calculate the chromatic aberration by treating ϵ as a perturbation parameter. If we combine this perturbation with that of (4.3.1), we obtain the expansion

$$\begin{aligned} \mathsf{P}m = {} & m^{(2)} + \epsilon m^{(2)\mathrm{I}} + \epsilon^2 m^{(2)\mathrm{II}} + \dots \\ & + \rho m^{(4)} + \epsilon\rho m^{(4)\mathrm{I}} + \dots \\ & + \dots . \end{aligned} \qquad (4.4.1)$$

‡ See, for instance, O. Scherzer, *J. Appl. Phys.* **20** (1949), 20–9.

§ See, for instance, G. Liebmann, *Proc. Phys. Soc.* B, **64** (1951), 972–7.

|| See, for instance, G. Liebmann and E. M. Grad, *Proc. Phys. Soc.* B, **64** (1951), 956–71; and G. Liebmann, *Proc. Phys. Soc.* B, **64** (1951), 972–7; **65** (1952), 94–108; and **65** (1952), 188–92.

‡ See refs. (29)–(32).

The term involving ϵ which is of the lowest order in ϵ and ρ is $\epsilon m^{(2)\mathrm{I}}$; this function will yield the most important contribution to the *paraxial chromatic aberration.*

If we introduce the characteristic function defined by

$$V_{ob}^{(2)\mathrm{I}} = \int_{z_o}^{z_b} m^{(2)\mathrm{I}}\, dz, \tag{4.4.2}$$

and expressed as $V_{ob}^{(2)\mathrm{I}}(\overline{w}_o, w_o, \overline{w}_a, w_a)$, we see from (3.5.25) that the paraxial chromatic aberration is given by $\epsilon w_b^{(1)\mathrm{I}}$, where

$$w_b^{(1)\mathrm{I}} = 2k^{-1} g_b \frac{\partial V_{ob}^{(2)\mathrm{I}}}{\partial \overline{w}_a}. \tag{4.4.3}$$

It is not difficult to see that this characteristic function must be expressible in the form

$$V_{ob}^{(2)\mathrm{I}} = \begin{pmatrix} \overline{w}_o \\ \overline{w}_a \end{pmatrix}' \begin{pmatrix} U^* & \bar{S}^* \\ S^* & T^* \end{pmatrix} \begin{pmatrix} w_o \\ w_a \end{pmatrix}, \tag{4.4.4}$$

where U^* and T^* are real and S^* may be complex if the field is partly or wholly magnetic. We may now deduce, with the help of (4.2.18) and (4.2.25), that

$$V_b^{(1)\mathrm{I}} = Sv_o + Tv_a, \tag{4.4.5}$$

where $$S = M.2k^{-1}S^*, \quad T = M.2k^{-1}\mathfrak{p}_o^{-\frac{1}{2}}\mathfrak{p}_a^{\frac{1}{2}}T^*. \tag{4.4.6}$$

We shall see that S represents a change of magnification while T represents defocusing.

The effect of a change of aperture position upon the coefficients of chromatic aberration may be calculated by the same technique as was adopted in the last section. We now obtain, in place of the relations (4.3.7), the relations

$$\tilde{S} = S - \mathfrak{p}_o^{\frac{1}{2}}\mathfrak{p}_{\tilde{a}}^{-\frac{1}{2}} g_{\tilde{a}} h_{\tilde{a}}^{-1} T, \quad \tilde{T} = \mathfrak{p}_{\tilde{a}}^{\frac{1}{2}}\mathfrak{p}_a^{-\frac{1}{2}} h_{\tilde{a}}^{-1} T, \tag{4.4.7}$$

from which it follows that S_r can always be eliminated merely by appropriate choice of the aperture position, but S_i and T can never be annulled in this way.

It is also possible to establish the geometrical significance of the coefficients S and T by reapplying the methods of the last section. We find, without difficulty, that the effect of S is to change the magnification coefficient from M to $M+\epsilon S$, which represents a change of magnification and also a rotation of the image if S is complex. Hence if ϵ takes a range of values between $-\boldsymbol{\epsilon}$ and $\boldsymbol{\epsilon}$, a point object is imaged as a line of length $2\epsilon|S|$ which is inclined at an angle θ_s to the sagittal direction. The coefficient T is found to represent a displacement of the focus by a distance

$$\epsilon M k^{-1} \mathsf{p}_0^{-\frac{1}{2}} \mathsf{p}_a^{-\frac{1}{2}} \mathsf{p}_b T,$$

but if, as before, ϵ takes a *range* of values, the effect of T upon the image is to change the image of a point object into a disk of radius $\epsilon T r_A$.

We see from (4.4.7) that the coefficient of chromatic defocusing is, in a sense, independent of aperture position, so that it would seem convenient to introduce a new coefficient so related to T that the radius of the chromatic aberration disk is given by $\epsilon r_b^{(1)\mathrm{I}}$, where

$$r_b^{(1)\mathrm{I}} = |MC_c|\,\omega_o \tag{4.4.8}$$

and, it will be remembered, ω_o is the semi-aperture angle of the beam at the object. We find that

$$C_c = 2k^{-2}T^* \tag{4.4.9}$$

and that, in terms of C_c, the displacement of the focus due to an increase ϵ in the energy is $\epsilon M^2 \mathsf{p}_0^{-1} \mathsf{p}_b C_c$. If the disk radius and shift of focus are referred back to the object plane, their values become $\epsilon|C_c|\,\omega_o$ and ϵC_c which are related exactly as one would expect.

In order to establish explicit formulae for the coefficients of chromatic aberration, it will be necessary to obtain the function $m^{(2)\mathrm{I}}$. This is simply the partial derivative of $m^{(2)}$—given by (4.2.12)‡—with respect to Φ, so that

$$m^{(2)\mathrm{I}} = \tfrac{1}{2}\mathsf{p}^{-1}(1+\Phi)\bar{u}'u' + \tfrac{1}{4}\mathsf{p}^{-3}\Phi''\bar{u}u. \tag{4.4.10}$$

If the transformations (4.2.15) and (4.2.18) are followed by partial integrations, this may be put into the form

$$m^{(2)\mathrm{I}} = \frac{\mathrm{d}}{\mathrm{d}z}(\mathscr{G}\Upsilon') + G\Upsilon + K\Omega, \tag{4.4.11}$$

where

$$\left.\begin{aligned} \mathscr{G} &= \tfrac{1}{4}\mathsf{p}^{-2}\langle 1+\Phi\rangle, \\ G &= \tfrac{1}{4}\{6\mathsf{p}^{-6}\langle 1+\tfrac{1}{6}\mathsf{p}^2\rangle\langle 1+\Phi\rangle\Phi'^2 + \mathsf{p}^{-4}\langle 1+\Phi\rangle \mathsf{H}^2\}, \\ K &= \tfrac{1}{4}\mathsf{p}^{-3}\langle 1+\Phi\rangle \mathsf{H}, \end{aligned}\right\} \tag{4.4.12}$$

‡ It is important that one should differentiate $m^{(2)}$ in the form (4.2.12) and *then* transform the co-ordinates, for if one differentiates $m^{(2)}$ given by (3.3.25) or (3.3.26), together with (4.2.17) or (4.2.19), the resulting estimates of chromatic aberration will contain a spurious term arising from the change of co-ordinates with beam energy.

and Υ and Ω are given by (4.3.25). We find from (4.2.21), (4.4.2) and (4.4.11) that the coefficients of (4.4.4) are related to the functions (4.4.12) by the formulae

$$\left.\begin{aligned} S^* &= -k[\mathscr{G}]_0^b + \int_{z_0}^{z_b} Ggh\,dz - ik\int_{z_0}^{z_b} K\,dz, \\ T^* &= \int_{z_0}^{z_b} Gh^2\,dz. \end{aligned}\right\} \tag{4.4.13}$$

It is obvious from (4.4.12) and (4.4.13) that T^* is essentially positive, so that, from (4.4.9), *the coefficient of defocusing C_c is essentially positive*. This is merely a mathematical formulation of the qualitative rule stated in § 1.5, that the paraxial image moves *away* from the object as the beam energy increases.

It follows from (4.4.9), (4.4.13) and (4.3.31) that

$$C_c = 2\int_{z_0}^{z_b} 2G\tilde{h}^2\,dz, \tag{4.4.14}$$

and it is not difficult to verify from this formula, (4.4.12), (1.4.31) and (1.5.14) that the non-relativistic thin-lens approximations to C_c, analogous to (4.3.32), are $2(z_a - z_o)^2 f^{-1}\Phi^{-1}$ and $(z_a - z_o)^2 f^{-1}\Phi^{-1}$ for electric and magnetic lenses, respectively. It is generally true that the chromatic aberration of an electric lens is about twice as great as that of a similar magnetic lens.

Let us now proceed to the *relativistic correction*, i.e. to a perturbation treatment of the difference between relativistic and non-relativistic calculations. If we take

$$p = \sqrt{(2\phi + \sigma\phi^2)} \tag{4.4.15}$$

instead of (1.2.4) as our initial relation between p and ϕ, we may obtain non-relativistic formulae by putting $\sigma = 0$ and relativistic formulae by putting $\sigma = 1$. If the beam energies are less than about 100,000 volts, i.e. $\phi < 0{\cdot}2$, terms which are linear in σ will give the principal contribution to the difference between the two treatments. We should therefore expand the variational function in the form

$$Pm = m^{(2)} + \sigma m^{(2)R} + \sigma^2 m^{(2)RR} + \dots, \tag{4.4.16}$$

and calculate the effect of the term $m^{(2)R}$, putting $\sigma = 1$.‡

It is clear from a comparison of (4.4.1) and (4.4.16) that the calculation follows the same lines as the calculation of the paraxial

‡ See p. 110, n. ‡.

chromatic aberration. We find, by substituting the formula (4.4.15) for p in (4.2.12), that

$$m^{(2)R} = \tfrac{1}{16}\mathsf{p}^3\bar{u}'u' - \tfrac{3}{32}\mathsf{p}\Phi''\bar{u}u. \qquad (4.4.17)$$

Hence we find that the paraxial relativistic correction in the image plane $v_b^{(1)R}$ is given by the right-hand side of (4.4.5), together with (4.4.6) and (4.4.13), if the formulae (4.4.12) are replaced by

$$\mathscr{G} = \tfrac{1}{32}\mathsf{p}^2, \quad G = \tfrac{1}{32}(2\mathsf{p}^{-2}\Phi'^2 + \mathsf{H}^2), \quad K = \tfrac{1}{32}\mathsf{pH}, \qquad (4.4.18)$$

where, it should be remembered, p is now given by its non-relativistic formula and the functions $g(z)$ and $h(z)$ which appear in (4.4.13) will be solutions of the non-relativistic form of the paraxial ray equation (4.2.20). It will be noted that G is again essentially positive.

The relativistic correction, as it has been calculated, may possibly be classed as an 'aberration', but it cannot be classed as an 'image defect', for the only change in the image is in its magnification and its focusing—which can be compensated. However, let us now consider a *purely electric* system comprising an electron gun and a lens system, the potentials of which are all derived from a common source. If the source potential fluctuates by a fraction v, ϕ and hence Φ, Φ', etc., will fluctuate by the same fraction. Now it is obvious from (1.3.4) that such fluctuations will have no optical significance if the beam energy is small enough for the non-relativistic treatment to be appropriate, but that, if a relativistic treatment is necessary, the optical characteristics of the system will be affected by such fluctuations. Since it is particularly convenient to derive the potentials of an all-electric system from a common source, it is interesting to investigate this 'relativistic aberration' or 'fluctuation aberration' which constitutes an image defect.

If ϕ is changed to $(1+v)\phi$, we may suppose p to be changed to $\sqrt{(2\phi + (1+v)\phi^2)}$, since the multiplying factor $\sqrt{(1+v)}$ is without significance. If this expression is substituted for p in (4.2.11), we may expand m as

$$Pm = m^{(2)} + vm^{(2)F} + v^2 m^{(2)FF} + \ldots, \qquad (4.4.19)$$

where, in particular,

$$m^{(2)F} = \tfrac{1}{4}\mathsf{p}^{-1}\Phi^2\bar{u}'u' - \tfrac{1}{8}\mathsf{p}^{-3}(3+\Phi)\Phi^2\Phi''\bar{u}u. \qquad (4.4.20)$$

Hence we find that the paraxial relativistic—or fluctuation—aberration in the image plane is given by $vv_b^{(1)F}$, where $v_b^{(1)F}$ may be obtained from (4.4.5) provided that the formulae (4.4.12) are replaced by

$$\mathscr{G}=\tfrac{1}{8}\mathrm{p}^{-2}\Phi^2,\quad G=\tfrac{1}{4}\mathrm{p}^{-6}\langle 1-\Phi\rangle\Phi^2\Phi'^2,\quad K=0. \tag{4.4.21}$$

It is interesting to note that in this case it is possible for T to vanish provided that Φ takes values above and below unity, i.e. provided that the beam energy takes values above and below 511,200 volts. This energy is, of course, much higher than that reached in the electrostatic electron microscope, so that the relativistic aberration cannot in practice be eliminated.

It is also worth noting that for values of the source potential V small enough for the non-relativistic form of (4.4.21) to be acceptable, the relativistic aberration varies with v and V as the combination vV, i.e. as ΔV, the fluctuation in V.

4.5. Space charge and space current

It was assumed in §4.2, when the variational function was calculated, that there were no charge or current distributions in the neighbourhood of the axis. However, the beam itself carries a charge distribution so that, if the beam is dense enough, the electric field will be modified by its presence. Moreover, if the beam energy is high enough for relativistic effects to be apparent, the current which the beam carries will affect the magnetic field.‡ The study of *controlled* space-charge distributions is also of some interest, since it is theoretically possible to eliminate, with their help, spherical aberration and the relativistic aberration (but not the chromatic aberration).

Let us replace the equation (4.2.1) by (1.2.8), expanding ρ in the form

$$\rho=\rho_0-\tfrac{1}{4}r^2\rho_2+\dots. \tag{4.5.1}$$

Then we obtain, in place of the expansion (4.2.2),

$$\phi=\Phi-\tfrac{1}{4}\bar{u}u(\Phi''+\rho_0)+\tfrac{1}{64}\bar{u}^2u^2(\Phi^{\mathrm{iv}}+\rho_0''+\rho_2)-\dots. \tag{4.5.2}$$

In order to take account of space currents, we should note that, since the charge distribution is static,

$$\operatorname{div}\mathbf{j}=0, \tag{4.5.3}$$

‡ It is worth noticing that the electric field due to space charge may be neutralized (see p. 117), but not so the magnetic field due to space current.

so that the current density is derivable from a complex function J, analogous to U, according to the equations

$$j_x + \mathrm{i}j_y = -\frac{\partial J}{\partial z}, \quad j_z = \frac{\partial \bar{J}}{\partial \bar{u}} + \frac{\partial J}{\partial u}. \tag{4.5.4}$$

The field equation (1.2.9) now leads to the equation

$$\nabla^2 U = -J \tag{4.5.5}$$

in place of (4.2.8). Hence, if we expand J as

$$J = \mathrm{i}\{\tfrac{1}{2}uJ_1 - \tfrac{1}{16}\bar{u}u^2 J_3 + \ldots\}, \tag{4.5.6}$$

the expansion for U becomes

$$U = \mathrm{i}\{\tfrac{1}{2}u\Psi - \tfrac{1}{16}\bar{u}u^2(\Psi'' + J_1) + \tfrac{1}{192}\bar{u}^2u^3(\Psi^{\mathrm{iv}} + J_1'' + J_3) - \ldots. \tag{4.5.7}$$

Although it is convenient to use the expansion (4.5.7) in obtaining the variational function, it is necessary to relate the coefficients J_1, J_3, etc., to the components of the space-current vector. Let us expand j_z, j_r, j_θ as

$$\left.\begin{aligned} j_z &= j_{z0} - \tfrac{1}{4}r^2 j_{z2} + \ldots, \\ j_r &= \tfrac{1}{2}rj_{r1} - \tfrac{1}{16}r^3 j_{r3} + \ldots, \\ j_\theta &= \tfrac{1}{2}rj_{\theta 1} - \tfrac{1}{16}r^3 j_{\theta 3} + \ldots; \end{aligned}\right\} \tag{4.5.8}$$

then we find that the coefficients in (4.5.6) and (4.5.8) are related by

$$\left.\begin{aligned} j_{z0} &= -J_{1i}, & j_{z2} &= -J_{3i}, \\ j_{r1} &= J'_{1i}, & j_{r3} &= J'_{3i}, \\ j_{\theta 1} &= -J'_{1r}, & j_{\theta 3} &= -J'_{3r}, \end{aligned}\right\} \tag{4.5.9}$$

where $J_1 = J_{1r} + \mathrm{i}J_{1i}$, etc. We may also note that although it is not possible to deduce from (1.2.10) general relations between the 'off-axis' coefficients in the expansions (4.5.1) and (4.5.8), it is possible to relate the 'axial' coefficients by

$$j_{z0} = \mathsf{p}(1+\Phi)^{-1}\rho_0. \tag{4.5.10}$$

We may now replace the expansions (4.2.2) and (4.2.10) by (4.5.2) and (4.5.7) and repeat the steps of § 4.2. In place of (4.2.17) and (4.2.19) we obtain

$$\mathsf{V} = \tfrac{1}{4}\{2\mathsf{p}^{-1}\langle 1+\Phi\rangle(\Phi'' + \rho_0) + \mathsf{p}^{-1}\mathsf{H}^2 - 2j_{z0}\} \tag{4.5.11}$$

and

$$\mathsf{W} = \tfrac{1}{4}\{3\mathsf{p}^{-4}\langle 1 + \tfrac{1}{3}\mathsf{p}^2\rangle\Phi'^2 + 2\mathsf{p}^{-2}\langle 1+\Phi\rangle\rho_0 + \mathsf{p}^{-2}\mathsf{H}^2 - 2\mathsf{p}^{-1}j_{z0}\}. \tag{4.5.12}$$

If we are considering the intrinsic charge and current of the beam, the terms in ρ_0 and j_{z0} may be combined, by means of (4.5.10), in the form $2\mathrm{p}^{-2}\langle 1+\Phi\rangle^{-1}\rho_0$ or $2\mathrm{p}^{-3}j_{z0}$. The latter is probably the more useful.

Let us consider, as an example, the divergence of a narrow beam under the influence of its intrinsic space charge. If the radius of the beam is R and the total current $-I$,

$$\pi R^2 j_{z0} = -I, \tag{4.5.13}$$

so that the ray equation, which may be written as

$$\frac{d^2u}{dz^2} - \tfrac{1}{2}\mathrm{p}^{-3}j_{z0}u = 0, \tag{4.5.14}$$

if we ignore the variation of the beam energy on the axis due to the space charge, may be rewritten as

$$\frac{d^2u}{dz^2} = \frac{1}{2\pi}\mathrm{p}^{-3}IR^{-2}u. \tag{4.5.15}$$

Since this equation is linear, the radial charge distribution will remain uniform if it is initially uniform. The variation of the beam radius may be obtained by setting $u=R$ in (4.5.15). This equation may be integrated to give

$$z - z_m = \pm 2\pi^{\frac{1}{2}}\mathrm{p}^{\frac{3}{2}}I^{-\frac{1}{2}}R_m\int_0^{(\log(R/R_m))^{\frac{1}{2}}} e^{t^2}\,dt, \tag{4.5.16}$$

wherein z_m is the abscissa and R_m the radius of the 'neck' of the beam at which R is a minimum. As an indication of the magnitude of this effect, we may consider a beam of 1000 volts energy, so that $\Phi = 0{\cdot}00196$ e.o.u. and $\mathrm{p} = 0{\cdot}0626$ e.o.u., carrying a current of 1 mA., so that $I = 7{\cdot}37\times 10^{-7}$ e.o.u.; then we find that, if the beam is initially parallel and of radius 0·1 cm., its radius will have increased by almost 10 % in 2 cm.

Let us consider, as another example, the behaviour of a narrow beam whose space-charge forces are just balanced by an external magnetic field. Since the paraxial equation is linear, it is again possible for a uniform radial space-charge distribution to be maintained. If, in the equation

$$\frac{d^2r}{dz^2} + \tfrac{1}{4}(\mathrm{p}^{-2}\mathrm{H}^2 - 2\mathrm{p}^{-3}j_{z0})r = 0, \tag{4.5.17}$$

we put $r = R + \Delta R$, where R is supposed to be the stable radius and ΔR the deviation from this radius, and if we now use the relation

$$\pi(R+\Delta R)^2 j_{z0} = -I, \tag{4.5.18}$$

we obtain the relation $\quad \mathrm{H}^2R^2 = 2\pi^{-1}\mathrm{p}^{-1}I, \qquad (4.5.19)$

and the equation, valid to the first order in ΔR,

$$\frac{d^2\Delta R}{dz^2}+\tfrac{1}{2}p^{-2}H^2\Delta R=0. \qquad (4.5.20)$$

It is not difficult to verify that, in the example considered at the end of the previous paragraph, the beam would maintain a stable radius of 0·1 cm. in the presence of a field of 0·0274 e.o.u., i.e. 46·7 gauss, and that a small deviation from this radius would undulate with a wave-length of 20·9 cm.

Let us now consider the modifications in the formulae for the chromatic aberration and the relativistic aberration which are brought about by the presence of space charge and space current. We find that in the chromatic-aberration formulae (4.4.12) the formula for G is altered to

$$G=\tfrac{1}{4}\{6p^{-6}\langle 1+\tfrac{1}{6}p^2\rangle\langle 1+\Phi\rangle\Phi'^2+2p^{-4}\langle 1+\tfrac{1}{2}p^2\rangle\rho_0 + p^{-4}\langle 1+\Phi\rangle H^2-p^{-3}\langle 1+\Phi\rangle j_{z0}\}, \qquad (4.5.21)$$

and that the formula for G in (4.4.21) becomes

$$G=\tfrac{1}{4}\{p^{-6}\langle 1-\Phi\rangle\Phi^2\Phi'^2-p^{-4}\Phi^2\rho_0-\tfrac{1}{2}p^{-3}\Phi^2 j_{z0}\}. \qquad (4.5.22)$$

Now in order to eliminate the aperture-dependent part of these aberrations, it is necessary to make G identically zero or partly negative. It is not difficult to see, from a comparison of (4.5.21) and (4.5.22) with (4.5.12), that it is possible to eliminate the chromatic aberration with a negative space-charge distribution, but that the lens would then be non-refractive or divergent, whereas a positive space charge is required to eliminate the relativistic aberration and that this makes the lens more strongly convergent.

It would be possible to write down new formulae for the third-order aberrations in place of (4.3.26) and (4.3.27), taking account of space charge and space current. However, the expressions are cumbersome and not particularly enlightening, so it is proposed to restrict our attention to the extension of formula (4.3.32) for the non-relativistic thin-lens approximation to the spherical aberration coefficient which takes account of space charge. The new equation may be written as

$$C_s=\tfrac{1}{64}(z_a-z_o)^4\left\{p^{-4}\int_{-\infty}^{\infty}(5\Phi''^2+9\Phi''\rho_0+4\rho_0^2)\,dz + p^{-2}\int_{-\infty}^{\infty}(2H^2-\rho_2)\,dz\right\}, \qquad (4.5.23)$$

from which it is at once obvious that it is not possible to correct a purely magnetic lens by means of a uniform space-charge distribution, i.e. one for which $\rho_2 = 0$ but $\rho_0 \neq 0$. The integrand of the first integral can be made to vanish or become negative by choosing ρ_0 according to $\rho_0 = \alpha\Phi''$, where $-\frac{5}{4} \leqslant \alpha \leqslant -1$, but the resulting lens would then be divergent provided that the magnetic field is not strong enough to make the spherical aberration coefficient positive. Hence in order to eliminate the spherical aberration of a lens by means of an auxiliary space-charge distribution, it is necessary for ρ_2 to be positive, i.e. for the charge to become more dense, if it is negative, or less dense, if it is positive, with increasing radius.

In order to obtain an estimate of the space charge required to achieve correction, we might consider the possibility of using four auxiliary electron beams forming angles of 45° at their point of intersection where they cross at right angles the beam whose spherical aberration is to be corrected. If the auxiliary beams can be formed with 'hour-glass' cross-section, the charge which is traversed by an electron of the primary beam travelling parallel to the axis will increase with its radius. If the main beam has an energy of 50,000 volts and is focused by a bell-shaped magnetic field of 1700 gauss maximum strength and of half-height width 0·5 cm., the focal length will be about 2 cm. If the auxiliary beams have an energy of 1000 volts, and if the radius of curvature of the 'neck' of the hour-glass is 0·2 cm., we find that the beams must have a current density of about 200 amperes per sq.cm. in order to achieve correction.

In order to obtain reliable estimates of the effect upon a beam of its intrinsic space charge, it is necessary to take into account the relevant electrode system rather than assume—as we have done—that it is the potential on the axis which is unaffected by the space charge. It is also important to mention that space-charge calculations may be—and often are—rendered invalid through neglect of the tendency of electron beams to become 'neutralized' by the positive charge of ions of the residual gases. For further details the reader is referred to Pierce.‡

‡ Ref. (2). See also E. L. Ginzton and B. H. Wadia, *Proc. Inst. Radio Engrs*, **42** (1954), 1548–54.

4.6. Asymmetries of electron lenses

If an electron-optical system deviates slightly from rotational symmetry, its imaging properties will only approximate to those calculated earlier in this chapter. It is therefore important to calculate the effects of asymmetries of electron lenses in order to determine the tolerances which should be assigned to the machining of the component parts of lenses and to the alinement of these parts.

The distortion of an electrode or pole-piece is determined by the displacement of points of its surface from the surface of its rotationally symmetrical ideal. Since this displacement is a function of the azimuthal angle θ, the distortion may be analysed by expressing this function as a Fourier series; we shall find that, if all the terms of this series were of comparable magnitudes, the most important would be those in $e^{\pm 2i\theta}$, which represent *ellipticity*. It is obvious that any misalinement of the components of a lens is an asymmetry of period 2π in θ.

Let us consider the characteristic function which represents the aberrations of an asymmetrical electron-optical system which, we suppose for simplicity, is imaging a point object at $w_o = 0$. Since the spherical aberration of a rotationally symmetrical system is non-zero, the principal aberrations must be accounted for if we include terms of up to the fourth order in the off-axis co-ordinates; hence we may write

$$V_{ob}^{(A)} = \begin{pmatrix} 1 \\ \bar{w}_a \\ \bar{w}_a^2 \end{pmatrix}' \begin{pmatrix} K^* & \bar{R}^* & \bar{Q}^* \\ R^* & L^* & \bar{P}^* \\ Q^* & P^* & N^* \end{pmatrix} \begin{pmatrix} 1 \\ w_a \\ w_a^2 \end{pmatrix}. \tag{4.6.1}$$

There is no point in including the terms in $\bar{w}_a w_a^3$ and $\bar{w}_a^4$ for, since these are introduced by asymmetries, they must be negligible compared with the spherical aberration term $N^* \bar{w}_a^2 w_a^2$; the absence of the term in $\bar{w}_a^3$ will be justified later. Only the coefficient N^* of (4.6.1) is identical with its counterpart in (4.3.4). We may deduce from (4.6.1) that

$$v_b^{(A)} = \begin{pmatrix} 1 \\ 2\bar{v}_a \end{pmatrix}' \begin{pmatrix} R & L & \bar{P} \\ Q & P & N \end{pmatrix} \begin{pmatrix} 1 \\ v_a \\ v_a^2 \end{pmatrix}, \tag{4.6.2}$$

where L, N, P, Q and R are related to L^*, N^*, P^*, Q^* and R^* by equations identical with (4.3.6). It is seen that P and Q still represent coma and astigmatism, but L and R represent simply defocusing and a shift of the image, both of which may be ignored.

If we now introduce three parameters α, β and γ which respectively characterize the magnitudes of the beam aperture, the distortion of the lens components, and the misalinement of these components, it will be possible to establish the principal aberrations arising from asymmetries by establishing the dependence of the terms involving P, Q and N on these parameters. The term $N\bar{v}_a v_a^2$ is known to behave as α^3; the term $P\bar{v}_a v_a$, since it arises from a term of (4.6.1) which is of period 2π, may contain components behaving as $\alpha^2\beta$ and as $\alpha^2\gamma$; similarly, since the term $Q^*\bar{w}_a^2$ of (4.6.1) is of period π, the term $2Q\bar{v}_a$ of (4.6.2) may contain components behaving as $\alpha\beta$ and as $\alpha\gamma^2$. We may now say that if the principal aberrations due to asymmetry are to be comparable with the spherical aberration, either $\alpha^2\beta$ or $\alpha\beta$ and either $\alpha^2\gamma$ or $\alpha\gamma^2$ should be of the same order as α^3, none being of lower order than α^3. Hence we should specify that $\beta = O(\alpha^2)$ and $\gamma = O(\alpha)$. Since the contributions to (4.6.2) arising from a term in $\bar{w}_a^3$ in (4.6.1) behaves as $\alpha^2\beta$ and $\alpha^2\gamma^3$, they may justifiably be ignored. It follows immediately that *the principal aberration due to distortion is astigmatism which is due to ellipticity of the pole-pieces or electrodes, varying linearly with the degree of ellipticity*, whereas *the principal aberrations due to misalinement are coma and astigmatism which vary linearly and quadratically, respectively, with the degree of misalinement.*

In order to evaluate the aberrations of an electron lens due to prescribed asymmetries, it is necessary first to calculate the resulting field in the neighbourhood of the axis and secondly to calculate the electron-optical properties of this field. A thorough treatment of this problem is lengthy,‡ but it is possible to give the calculation of an important special case for which the first of the above steps need not be considered in detail.

It has been found‡ that if the electrodes of an electrostatic lens or the pole-pieces of a magnetic lens suffer a 'uniform' elliptic distortion, the directions of the principal axes and the degree of ellipticity η being independent of z, the equipotential surfaces of the scalar potential in the neighbourhood of the axis are of elliptical cross-section and have approximately the same ellipticity.

‡ See refs. (34) and (35).

The general solution of the Laplace equation (4.2.1) may be written as

$$\phi = \sum_{h,l=0}^{\infty} \frac{(-)^l}{2^{h+l}h!\,l!} \Phi_{h-l}^{(2l)} \bar{u}^h u^l, \tag{4.6.3}$$

where the superfix to $\Phi_h(z)$ denotes differentiation with respect to z and $\Phi_{-h} = \bar{\Phi}_h$, but if we ignore terms of higher than the second order in u and suppose that $\Phi_1 \equiv 0$, the expansion reduces to

$$\phi = \Phi_0 - \tfrac{1}{4}\Phi_0'' \bar{u}u + \tfrac{1}{8}(\Phi_2 \bar{u}^2 + \bar{\Phi}_2 u^2). \tag{4.6.4}$$

We find that the equipotential surfaces are of the form

$$r = r_0(z)\,(1 + \eta \cos 2\theta) \tag{4.6.5}$$

if

$$\Phi_2 = 2\eta\Phi_0''. \tag{4.6.6}$$

The equations (4.6.4) and (4.6.6) may also serve as an expansion for the magnetic scalar potential ψ of an elliptical magnetic lens in terms of Ψ_0. Now in consequence of (4.2.9) the complex potential may be related to the coefficients of the scalar-potential expansion by the expansion

$$U = \sum_{h \geq l \geq 0} \frac{\mathrm{i}(-)^l}{2^{h+l}(h+1)!\,l!} \bar{\Psi}_{h-l}^{(2l)} u^{h+1} \bar{u}^l, \tag{4.6.7}$$

so that we may deduce from (4.6.4) and (4.6.6) that

$$U = \mathrm{i}(\Psi_0 u - \tfrac{1}{8}\Psi_0'' \bar{u}u^2 + \tfrac{1}{24}\bar{\Psi}_2 u^3), \tag{4.6.8}$$

where

$$\Psi_2 = 2\eta\Psi_0''. \tag{4.6.9}$$

There is no difficulty in recalculating the paraxial variational function from (4.2.11), replacing (4.2.2) and (4.2.10) by (4.6.4) and (4.6.8). We find that

$$m^{(2)} = m^{(G)} + m^{(A)}, \tag{4.6.10}$$

where $m^{(G)}$ is given by (4.2.12) and would, by itself, give Gaussian imaging, and $m^{(A)}$, the part due to asymmetries, is given by

$$m^{(A)} = \tfrac{1}{8}\mathrm{p}^{-1}\langle 1 + \Phi\rangle (\Phi_2 \bar{u}^2 + \bar{\Phi}_2 u^2) + \tfrac{1}{8}\mathrm{i}(\Psi_2 \bar{u}^2 - \bar{\Psi}_2 u^2). \tag{4.6.11}$$

The astigmatism coefficient Q^* may now be obtained in the usual way by carrying out the transformations (4.2.15) and (4.2.18) and then expressing the characteristic function (4.6.1) as the integral of the function (4.6.11). By restricting ourselves to purely electric and purely magnetic lenses we obtain the formulae

$$Q^* = \tfrac{1}{8}\int_{z_0}^{z_b} \mathrm{p}^{-2}\langle 1 + \Phi\rangle \Phi_2 h^2 \,\mathrm{d}z \tag{4.6.12}$$

and

$$Q^* = \tfrac{1}{8}\mathrm{i}\int_{z_0}^{z_b} \mathrm{p}^{-1}\Psi_2 \mathrm{e}^{-2\mathrm{i}\chi}\,\mathrm{d}z, \tag{4.6.13}$$

respectively.

Let us now assume that the lenses are weak and that the apertures are situated at the lenses. By using (4.6.6) and (4.6.9) and performing partial integrations, we obtain

$$Q^* = \tfrac{1}{2}\eta \mathrm{p}^{-4} \int_{-\infty}^{\infty} \Phi'^2 \, dz \qquad (4.6.14)$$

and

$$Q^* = \tfrac{1}{4}\eta \mathrm{p}^{-2} \int_{-\infty}^{\infty} \mathrm{H}^2 \, dz, \qquad (4.6.15)$$

where we have adopted the non-relativistic approximation in (4.6.14). By comparing these formulae with (1.4.31) and (1.5.14) it is possible to obtain approximate relations between the astigmatism coefficients and the focal length. The coefficient Q of (4.6.2) is related to Q^* by

$$Q = 2Mk^{-1}\mathrm{p}_0^{-\frac{1}{2}}\mathrm{p}_a^{\frac{1}{2}}Q^*, \qquad (4.6.16)$$

so we find that Q is given approximately by $-\frac{4}{3}\eta M$ and $-2\eta M$ for electric and magnetic lenses, respectively, in the important special case that the image plane is effectively at infinity.

We have noted in §§4.3 and 4.4 that it is possible to describe aberrations which are independent of the object co-ordinates in a form which is independent of the aperture position. It is seen from (4.6.1) and (4.6.2) that the astigmatism due to asymmetries is independent of the object co-ordinates, so that, by analogy with (4.4.8), we are led to express the aberration as

$$r_b^{(A)} = \eta M C_A \omega_o, \qquad (4.6.17)$$

where $r_b^{(A)}$ is the radius of the disk of least confusion, ω_o is the semi-aperture angle of the beam at the object, and C_A is an astigmatism coefficient which is related to Q^* by

$$C_A = 4k^{-2}\left|\frac{\partial Q^*}{\partial \eta}\right|. \qquad (4.6.18)$$

The astigmatic difference is given by

$$D_b^{(A)} = 2\eta \mathrm{p}_0^{-1}\mathrm{p}_b M^2 C_A, \qquad (4.6.19)$$

but if the aberration is referred back to the object plane, we obtain the simple formulae $\eta C_A \omega_o$ and $2\eta C_A$ for the radius of the disk of least confusion and the astigmatic difference, respectively. On reverting to the thin-lens approximation and assuming that the image is at infinity, we find that C_A is either $\frac{8}{3}f$ or $4f$, according as the lens is electric or magnetic.

If, as an example, we consider a magnetic objective lens of focal length 1 cm. operating at a semi-aperture angle of 0·003 radian, we find that if the diameter of the disk of least confusion is to be less than 100 Ångström units, i.e. 10^{-6} cm., the degree of ellipticity must be less than 4×10^{-5}. This means that, if the radius of the pole-pieces is 0·25 cm., the overall variation in radius must be less than 2×10^{-5} cm., i.e. 0·2 micron. Since it is impossible to meet such tolerances, very high resolutions have been attained in electron microscopy by introducing controlled asymmetrical fields to compensate for the lens asymmetries.‡

‡ See J. Hillier and E. G. Ramberg, *J. Appl. Phys.* **18** (1947), 48–71.

CHAPTER 5

SYSTEMS OF MIRROR SYMMETRY

5.1. Introduction

We have already seen in § 3.3 that it is possible for a system which is not of rotational symmetry to display paraxial image-forming properties provided the orthogonality condition (3.3.13) is satisfied. In this chapter we shall investigate a class of electron-optical systems with curved ray axes, the symmetry of which entails the fulfilment of the orthogonality condition; this class comprises systems of mirror symmetry.

An electron-optical system will be said to have 'mirror' symmetry if there exists a plane with respect to which the electric and magnetic scalar potentials have 'even' and 'odd' symmetry, respectively. This definition implies that, at the plane of symmetry, the electric and magnetic fields are parallel and normal, respectively, to this plane; an immediate consequence is that any electron ray which touches the plane of symmetry lies wholly in it, for an electron moving in such a trajectory will experience no force normal to the plane. In particular, the ray axis will be chosen to lie in the plane of symmetry.

Since the orthogonality condition is satisfied, the paraxial imaging properties of systems of mirror symmetry will differ only slightly from those of systems of rotational symmetry. However, we may infer from (3.3.1) that, since the variational function is no longer even in both off-axis co-ordinates, the most important geometrical aberrations will be of the second order in these co-ordinates. It will not be possible to devote space to an investigation of these aberrations.

The most important difference between systems of rotational symmetry and of mirror symmetry is in their chromatic aberration. We saw in the preceding chapter that the principal effects of a variation in beam energy were a defocusing and a change of magnification; similar, but slightly more complicated, effects appear in systems of mirror symmetry, but these also will be outside the scope of this chapter. The reason for this will be established in § 5.2: the calculation of the paraxial chromatic aberration is now a *second-order* perturbation problem which must be based upon a calculation of the second-order (geometrical) aberrations. The principal effect of a variation of beam energy, i.e. the effect which is of the lowest order in this variation and in the off-axis co-ordinates, may be termed the 'zero-order chromatic aberration' since it does not involve the off-axis co-ordinates; it represents a pure displacement of the image. This effect is, of course, to be expected from the fact that the ray axis varies with the beam energy and field strengths.

The zero-order chromatic aberration is more commonly referred to as 'dispersion'—or, more specifically, as 'energy dispersion'—by analogy with the dispersion of light-optical prisms. Just as a glass prism in a light-optical spectrometer is used to resolve a beam of light into its colour components, so an electric or magnetic field of mirror symmetry may act as a prism in resolving a beam of electrons into its energy components; if the electrons are those emitted by a radioactive source, the electron-optical system is known as a 'β-ray spectrometer'. The term 'β-ray spectrometer' is used in a restricted sense to denote an instrument in which the electrons emitted by a source, and suitably defined by a baffle or stop, are focused upon the window of a Geiger-Müller counter; the term 'β-ray spectrograph' is applied to an instrument in which electrons of any definite energy in a finite band of energies will produce a sharp focus of a line source upon a photographic plate. Since one examines the focusing properties of a β-ray spectrometer by constructing the image formed in the plane of the counter window, a study of the optical properties of β-ray spectrographs includes that of β-ray spectrometers. Examples of simple β-ray spectrographs will be given in § 5.4.

The effect of paraxial chromatic aberration is more complicated in systems of mirror symmetry than in rotationally symmetrical systems. The aperture-dependent part of this aberration is generally astigmatic, but it is always possible to adjust the field so that the aberration should not be astigmatic.‡ The principal effect of the paraxial chromatic aberration would then be a longitudinal (i.e. an

‡ See ref. (37), p. 185, where it is also demonstrated that a suitable choice of aperture position will make this aberration stigmatic.

'axial') shift of focus; it follows that this aberration, in conjunction with the zero-order chromatic aberration which produces a transverse shift of focus, determines the correct inclination (to the axis) of the image plane in a system, such as a β-ray spectrograph, which is to focus a band of energies.

Our adoption of electron-optical units has tended to obscure the dependence of the electron-optical properties of a system upon the specific charge η, where $\eta = e/m_0$, of the particles which constitute the beam. However, it will be remembered that the units of electric and magnetic potentials are proportional to η, so that η appears implicitly in the variational equation (1.3.4). We may infer that a system of mirror symmetry may be used to resolve a beam containing different ions into its components; such an instrument is known as a mass spectrometer or mass spectrograph, for, since the charge is known to be a small multiple of the electronic charge, the specific charge determines the mass of the ion. Since the units appropriate to ions are very large, relativistic effects are never important in mass spectrometry; it follows at once from the non-relativistic form of (1.3.4) that a purely electric system cannot resolve specific charges.

If a beam of ions were known to be of a definite energy, resolution could be obtained by means of a magnetic field alone. In order to achieve precision, however, it is necessary that the specific charge should be determined only by the field strengths of the deflecting fields which constitute the mass spectrometer. Hence modern instruments are of the 'double-focusing' variety which, for given specific charge, form a sharp image of an entrance slit not only for a finite aperture but also for a band of beam energy. An example of a simple mass spectrograph will be given in § 5.5.

5.2. The variational functions

The co-ordinate system will be based upon a ray axis which will, in the first instance, be supposed to be an arbitrary curve lying in the plane of symmetry. If the orthogonal co-ordinate system (x, y, z) is defined as in fig. 5.1 so that z measures arc length along the ray axis and the x- and y-axes are respectively contained in and normal to the plane of symmetry for any value of z, then the metric is given by

$$ds^2 = dx^2 + dy^2 + (1 - \kappa x)^2 dz^2, \qquad (5.2.1)$$

where $\kappa(z)$ is the curvature of the ray axis. It follows from (5.2.1) that h_x, h_y, h_z, which are the familiar coefficients associated with orthogonal curvilinear co-ordinates,‡ are given by

$$h_x = 1, \quad h_y = 1, \quad h_z = 1 - \kappa x. \tag{5.2.2}$$

Now since the electric scalar potential is even in y it may be expanded as

$$\phi = \Phi + \Phi_x x + \tfrac{1}{2}\Phi_{xx} x^2 + \Phi_{yy} y^2 + \ldots, \tag{5.2.3}$$

where $\Phi = \Phi(z)$, etc. We find from the Laplace equation§

$$\nabla^2\phi \equiv \frac{1}{h_x h_y h_z}\left\{\frac{\partial}{\partial x}\left(\frac{h_y h_z}{h_x}\frac{\partial\phi}{\partial x}\right) + \frac{\partial}{\partial y}\left(\frac{h_z h_x}{h_y}\frac{\partial\phi}{\partial y}\right) + \frac{\partial}{\partial z}\left(\frac{h_x h_y}{h_z}\frac{\partial\phi}{\partial z}\right)\right\} = 0 \tag{5.2.4}$$

that

$$\Phi_{yy} = -\Phi'' + \kappa\Phi_x - \Phi_{xx}. \tag{5.2.5}$$

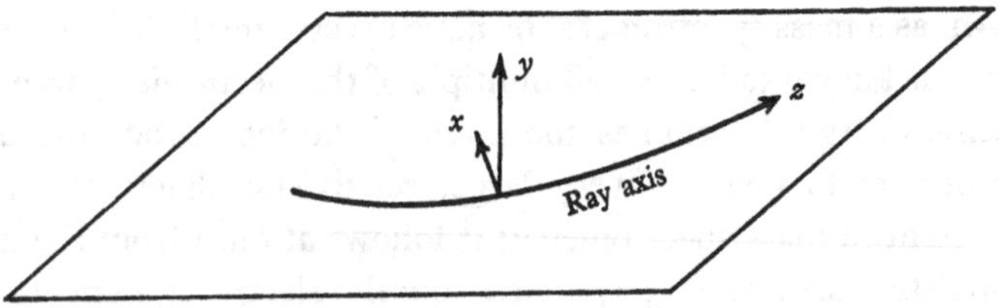

Fig. 5.1. Choice of axes for systems of mirror symmetry.

Since the magnetic scalar potential is antisymmetric in y, the vector potential should be expanded as

$$\left.\begin{aligned} A_x &= \mathsf{A}_x + \mathsf{A}_{x,x} x + \ldots, \\ A_y &= \mathsf{A}_{y,y} y + \ldots, \\ A_z &= \mathsf{A}_z + \mathsf{A}_{z,x} x + \tfrac{1}{2}\mathsf{A}_{z,xx} x^2 + \tfrac{1}{2}\mathsf{A}_{z,yy} y^2 + \ldots, \end{aligned}\right\} \tag{5.2.6}$$

and the field vector as

$$H_x = \mathsf{H}_{x,y} y + \ldots, \quad H_y = \mathsf{H}_y + \mathsf{H}_{y,x} x + \ldots, \quad H_z = \mathsf{H}_{z,y} y + \ldots, \tag{5.2.7}$$

The field vector and vector potential are related in curvilinear co-ordinates by‖

$$H_x = \frac{1}{h_y h_z}\left\{\frac{\partial}{\partial y}(h_z A_z) - \frac{\partial}{\partial z}(h_y A_y)\right\}, \quad \text{etc.,} \tag{5.2.8}$$

‡ See, for instance, J. C. Slater and N. H. Frank, *Electromagnetism* (McGraw-Hill, New York, 1947), Appendix 4.

§ Slater and Frank, loc. cit. p. 222, eq. (7).

‖ Slater and Frank, loc. cit. p. 223, eq. (11).

so that the coefficients appearing in (5.2.6) and (5.2.7) are related by

$$\left.\begin{aligned}\mathsf{H}_{x,y}&=\mathsf{A}_{z,yy}-\mathsf{A}'_{y,y}, \quad \mathsf{H}_y=\mathsf{A}'_x-\mathsf{A}_{z,x}+\kappa\mathsf{A}_z,\\ \mathsf{H}_{y,x}&=\mathsf{A}'_{x,x}+\kappa\mathsf{A}'_x-\mathsf{A}_{z,xx}+\kappa\mathsf{A}_{z,x}+\kappa^2\mathsf{A}_z.\end{aligned}\right\} \tag{5.2.9}$$

We should also note from (5.2.8) that, since the curl of the field vector vanishes,

$$\mathsf{H}_{x,y}=\mathsf{H}_{y,x}. \tag{5.2.10}$$

The above relations will make it possible to express all the potential coefficients appearing in the first-order and paraxial variational functions in terms of Φ, Φ_x, Φ_{xx}, H_y and $\mathsf{H}_{y,x}$.

If the electrodes are close enough to the ray axis, their shape may be related to the coefficients Φ, Φ_x and Φ_{xx} by means of (5.2.3) and (5.2.5). Similarly, if the magnetic field is produced by iron pole-pieces whose permeability may be assumed infinite, their shape may be related to H_y and $\mathsf{H}_{y,x}$ by means of the following expansion for the scalar potential:

$$\psi=-\mathsf{H}_y y-\mathsf{H}_{y,x}xy-\ldots. \tag{5.2.11}$$

We may now proceed to evaluate the variational function. On combining (1.3.10) and (3.2.2) and noting from (5.2.1) that the components of the direction vector are given by

$$(l_x, l_y, l_z)=\left(\frac{dx}{ds}, \frac{dy}{ds}, (1-\kappa x)\frac{dz}{ds}\right), \tag{5.2.12}$$

we see that

$$m=p\frac{ds}{dz}-x'A_x-y'A_y-(1-\kappa x)A_z. \tag{5.2.13}$$

The variational function may now be expanded as a series of polynomials in the off-axis co-ordinates, of the form (3.3.1), by means of (1.2.4), (5.2.1), (5.2.3) and (5.2.6). On using the partial-integration rule and the relations (5.2.5), (5.2.9) and (5.2.10), we obtain the formulae

$$m^{(1)}=(-\mathsf{p}\kappa+\mathsf{p}^{-1}(1+\Phi)\Phi_x+\mathsf{H}_y)x \tag{5.2.14}$$

and

$$\begin{aligned}m^{(2)}&=\tfrac{1}{2}\mathsf{p}(x'^2+y'^2)\\&\quad+\tfrac{1}{2}(-2\mathsf{p}^{-1}\kappa(1+\Phi)\Phi_x-\mathsf{p}^{-3}\Phi_x^2+\mathsf{p}^{-1}(1+\Phi)\Phi_{xx}-\kappa\mathsf{H}_y+\mathsf{H}_{y,x})x^2\\&\quad+\tfrac{1}{2}(-\mathsf{p}^{-1}(1+\Phi)\Phi''+\mathsf{p}^{-1}\kappa(1+\Phi)\Phi_x-\mathsf{p}^{-1}(1+\Phi)\Phi_{xx}-\mathsf{H}_{y,x})y^2,\end{aligned} \tag{5.2.15}$$

where p is defined by (4.2.14).

It has already been observed in § 3.3 that the condition that the axis should be a ray is that $m^{(1)}$ should vanish identically. Hence we obtain from (5.2.14) the relation

$$\mathsf{p}\kappa - \mathsf{p}^{-1}(1+\Phi)\Phi_x - \mathsf{H}_y = 0, \tag{5.2.16}$$

which may be used to eliminate either Φ_x or H_y from (5.2.15). The paraxial variational function is seen to be in orthogonal form (cf. (3.3.4)), so that the axes we have chosen are principal axes. The paraxial approximations to the ray variables are found from (3.2.5) and (5.2.15) to be

$$p_x = \mathsf{p}x', \quad p_y = \mathsf{p}y'. \tag{5.2.17}$$

We should note that, since the x-y co-ordinates form a Cartesian set, p_x and p_y as given by (5.2.17) may be identified with the corresponding components of the ray vector.

There is no difficulty in obtaining the paraxial ray equations in the form (3.3.6) from (5.2.15), but this will be left for later sections. However, it is interesting to note that if, on passing to the notation of (3.3.4), the transformations (3.3.15) and (3.3.17) are carried out, the following relation holds:

$$\mathsf{Q}_1 + \mathsf{Q}_2 = \kappa^2 + \tfrac{1}{2}\mathsf{p}^{-4}(3+\mathsf{p}^2)\Phi'^2. \tag{5.2.18}$$

Since, in virtue of (5.2.18), Q_1 and Q_2 cannot both be negative, we see from (3.3.18) that, in a certain sense, *no system of mirror symmetry can be wholly divergent*; it is, of course, implicitly understood that the neighbourhood of the ray axis is free of space charge.

5.3. The aberration characteristic functions

It is now proposed that, before proceeding to the consideration of specific instruments, the formal problem of the relation of aberration characteristic functions to the aberration terms of the variational function should be considered. It is important to see why the paraxial chromatic aberration cannot be obtained by purely paraxial calculations if we are to avoid a prevalent error.

There is no difficulty in seeing how to calculate the geometrical aberrations due to the terms $m^{(3)}$, etc., of (3.3.1). One may introduce, as in (4.3.1), a fictitious perturbation parameter and write

$$Pm = m^{(2)} + \rho m^{(3)} + \rho^2 m^{(4)} + \ldots. \tag{5.3.1}$$

Since our work on perturbation characteristic functions was carried only to the first order, we can write down formulae only for the aberrations arising from $m^{(3)}$, namely, the second-order aberrations. If, following (3.5.6), we write

$$V_{ob}^{(3)} = \int_{z_o}^{z_b} m^{(3)}\,dz, \tag{5.3.2}$$

it follows from (3.5.21), on making the obvious appropriate changes in the notation, that

$$x_b^{(2)} = k_x^{-1} g_{xb} \frac{\partial V_{ob}^{(3)}}{\partial x_a}, \quad y_b^{(2)} = k_y^{-1} g_{yb} \frac{\partial V_{ob}^{(3)}}{\partial y_a}. \tag{5.3.3}$$

The difficulty appears when we introduce a second perturbation such as that due to a change in beam energy, for terms derived from $m^{(1)}$ are then introduced. The formula, which is precisely analogous to (4.4.1), would include terms such as $\epsilon\rho^{-1}m^{(1)\mathrm{I}}$, which clearly cannot be treated by the method of § 3.5. However, let us suppose that in place of ϵ we write $\epsilon\rho$; this will not affect the significance of ϵ, since ρ is supposed to take the value unity; nor, for each power of ϵ, will it affect the separation of terms according to their powers as polynomials in the off-axis co-ordinates. This is, therefore, an acceptable choice of perturbation parameters which enables us to write

$$\begin{aligned} Pm = \quad & \epsilon m^{(1)\mathrm{I}} + \epsilon^2 \rho m^{(1)\mathrm{II}} + \dots \\ & + m^{(2)} + \epsilon\rho m^{(2)\mathrm{I}} + \dots \\ & + \rho m^{(3)} + \epsilon\rho^2 m^{(3)\mathrm{I}} + \dots \\ & + \dots . \end{aligned} \tag{5.3.4}$$

It is seen that, apart from $m^{(3)}$, the only first-order term appearing in the perturbation expansion (5.3.4) is $m^{(1)\mathrm{I}}$, which is responsible for the 'zero-order chromatic aberration' or 'dispersion'; this is reasonable for, of all the terms associated with the first power of ϵ, $m^{(1)\mathrm{I}}$ is of the lowest order in the off-axis co-ordinates. If we define the characteristic function

$$V_{ob}^{(1)\mathrm{I}} = \int_{z_o}^{z_b} m^{(1)\mathrm{I}}\,dz, \tag{5.3.5}$$

the zero-order chromatic aberration is given by

$$x_b^{(0)\mathrm{I}} = k_x^{-1} g_{xb} \frac{\partial V_{ob}^{(1)\mathrm{I}}}{\partial x_a}, \quad y_b^{(0)\mathrm{I}} = k_y^{-1} g_{yb} \frac{\partial V_{ob}^{(1)\mathrm{I}}}{\partial y_a}. \tag{5.3.6}$$

It is obvious that since this aberration—being of zero order in the off-axis co-ordinates—is independent of the object and aperture co-ordinates, it represents merely a displacement of the image.

It is from the term $m^{(2)\mathrm{I}}$ of (5.3.4) that the paraxial chromatic aberration—which, we saw in §4.4, is the principal chromatic aberration of rotationally symmetrical systems—must be calculated. Since this term is of the second order in the perturbation parameters, it is impossible to calculate the defocusing and change of magnification due to a change in energy without performing a second-order perturbation calculation which we are not in a position to do.‡ Even if the relevant formulae were invoked, the calculation would still be outside the scope of this chapter since, as might be expected from examination of (5.3.4), the evaluation of the aberration due to $m^{(2)\mathrm{I}}$ must take into account the aberrations due to $m^{(1)\mathrm{I}}$ and $m^{(3)}$, the latter of which we are not proposing to evaluate.

5.4. β-ray spectrographs

In this section we shall consider a few simple electron-optical systems which constitute possible β-ray spectrographs. If the field is purely electric, Φ_x may be eliminated from (5.2.15) by means of (5.2.16); the paraxial ray equations are then found to be

$$\left.\begin{aligned}\frac{d}{dz}\left(\mathrm{p}\frac{dx}{dz}\right)+\left(3\mathrm{p}\kappa^2\langle 1+\tfrac{2}{3}\mathrm{p}^2\rangle\langle 1+\Phi\rangle^{-2}-\mathrm{p}^{-1}\langle 1+\Phi\rangle\Phi_{xx}\right)x=0,\\ \frac{d}{dz}\left(\mathrm{p}\frac{dy}{dz}\right)+\left(-\mathrm{p}\kappa^2+\mathrm{p}^{-1}\langle 1+\Phi\rangle\Phi''+\mathrm{p}^{-1}\langle 1+\Phi\rangle\Phi_{xx}\right)y=0,\end{aligned}\right\} \tag{5.4.1}$$

and the function $m^{(1)\mathrm{I}}$ is found from (5.2.14) and (5.2.16) to be

$$m^{(1)\mathrm{I}}=-2\mathrm{p}^{-1}\langle 1+\tfrac{1}{2}\mathrm{p}^2\rangle\kappa\langle 1+\Phi\rangle^{-1}x. \tag{5.4.2}$$

It will be remembered that terms enclosed in angular brackets are to be replaced by unity in non-relativistic calculations, and that $m^{(1)\mathrm{I}}$ is found simply by taking the derivative of $m^{(1)}$ with respect to Φ. It is obvious from (5.4.1) that, if relativistic effects are taken into account, the focusing properties of the system will vary with the beam energy so that purely electric β-ray spectrographs can be used only for 'non-relativistic' energies, i.e. for energies less than about 50,000 volts. We shall therefore adopt the non-relativistic approximation.

It is further proposed that, for simplicity, we assume that κ, Φ, Φ_x and Φ_{xx} are all independent of z; the ray axis will then be a circle of

‡ For further details, see refs. (25) and (37).

radius a, where $a = \kappa^{-1}$, and the field will have the symmetry of a transverse section of an axially symmetrical field. Such fields are known as 'sector fields' if they extend only partly over the ray axis. The equations (5.4.1) may now be written as

$$\left.\begin{aligned} \frac{d^2x}{dz^2} + (3\kappa^2 - p^{-2}\Phi_{xx})\,x = 0, \\ \frac{d^2y}{dz^2} + (-\kappa^2 + p^{-2}\Phi_{xx})\,y = 0. \end{aligned}\right\} \quad (5.4.3)$$

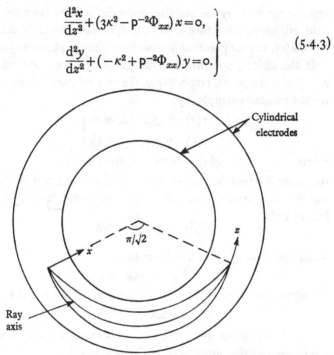

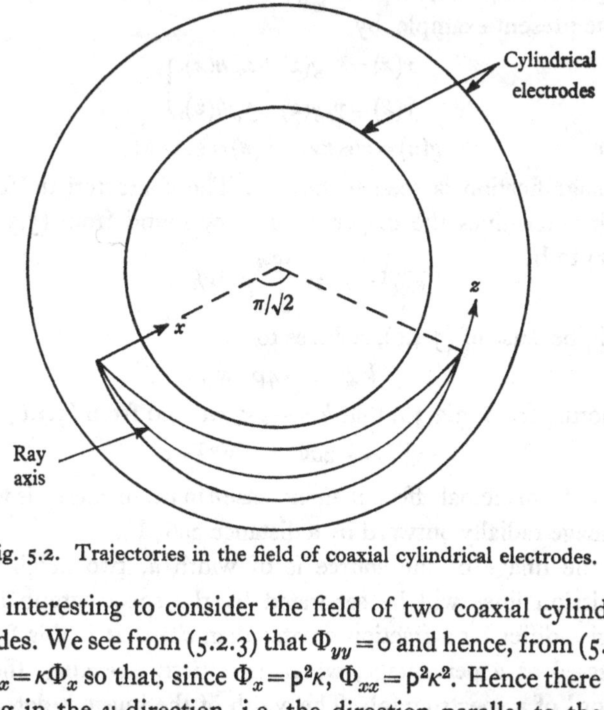

Fig. 5.2. Trajectories in the field of coaxial cylindrical electrodes.

It is interesting to consider the field of two coaxial cylindrical electrodes. We see from (5.2.3) that $\Phi_{yy} = 0$ and hence, from (5.2.5), that $\Phi_{xx} = \kappa\Phi_x$ so that, since $\Phi_x = p^2\kappa$, $\Phi_{xx} = p^2\kappa^2$. Hence there is no focusing in the y-direction, i.e. the direction parallel to the axis of the cylinders, but rays which lie in the plane of symmetry cut the ray axis at intervals $\pi/(\kappa\sqrt{2})$, i.e. $(\pi/\sqrt{2})\,a$ (fig. 5.2).

The field of coaxial cylinders could be used as a β-ray spectrograph since it is imperative to secure sharp imaging only in the direction of the chromatic displacement which will obviously be the x-direction. However, one may obtain a brighter image of the emitter by arranging to secure stigmatic imaging. This is obtained if $\Phi_{xx} = 2p^2\kappa^2$, so that $\Phi_{xx} = 2\kappa\Phi_x$, and points of the ray axis which are diametrically opposite are then conjugate points. We find from (5.2.3), (5.2.5) and (5.2.16) that the potential of this arrangement is given by the expansion

$$\phi = \Phi(1 + 2\kappa x + 2\kappa^2 x^2 - \kappa^2 y^2 + \dots), \tag{5.4.4}$$

from which we may see that the electrodes required to produce this field will have a curvature of $(a+2x)^{-1}$, directed towards the axis of symmetry, at their points of intersection with the plane of symmetry.

If the object, aperture and image planes are taken at $z_o = 0$, $z_a = \frac{1}{2}\pi a$ and $z_b = \pi a$, respectively, the general paraxial ray is given, for the present example, by

$$\left.\begin{aligned} x(z) &= x_o g(z) + x_a h(z), \\ y(z) &= y_o g(z) + y_a h(z), \end{aligned}\right\} \tag{5.4.5}$$

where $$g(z) = \cos \kappa z, \quad h(z) = \sin \kappa z; \tag{5.4.6}$$

the magnification is seen to be -1. The characteristic function which determines the dispersion is now found from (5.3.5) and (5.4.2) to be

$$V_{ob}^{(1)\mathrm{I}} = -2x_a \int_{z_o}^{z_b} p^{-1} \kappa h \, dz \tag{5.4.7}$$

which, because of (5.4.6), reduces to

$$V_{ob}^{(1)\mathrm{I}} = -4\mathrm{p}^{-1} x_a. \tag{5.4.8}$$

On noting from (3.3.36) that $k = -\mathrm{p}\kappa$, we find from (5.3.6) that

$$x_b^{(0)\mathrm{I}} = -2a\Phi^{-1}, \quad y_b^{(0)\mathrm{I}} = 0. \tag{5.4.9}$$

Hence the principal effect of an increase ϵ in beam energy is to move the image radially outward by a distance $2a(\epsilon/\Phi)$.

If the image of the source is of width d, two neighbouring β-emission lines will be separated by the spectrograph if their energies differ by a fraction greater than $d/2a$. It is this fraction, expressed as a percentage, which is, strictly speaking, the 'dispersion' of a spectrograph. The width of the image is determined not merely by the width of the source but also by the aberrations so that d increases with both the source size and the solid angle over which electrons are accepted. It is not possible to establish the important relation between the resolution and transmission of an instrument, on which the instrument is judged, without investigating the second-order aberrations. Nor is it possible to establish the best inclination of the image plane to the ray axis without calculating the paraxial chromatic aberration.

Since electrostatic β-ray spectrographs are suitable only for low energies, most instruments are purely magnetic. In this case the ray equations derivable from (5.2.15) and (5.2.16) are

$$\left.\begin{aligned}\frac{d^2x}{dz^2}+(\kappa^2-\mathsf{p}^{-1}\mathsf{H}_{y,x})\,x=0,\\ \frac{d^2y}{dz^2}+\mathsf{p}^{-1}\mathsf{H}_{y,x}y=0.\end{aligned}\right\} \tag{5.4.10}$$

It is clear that in a uniform magnetic field, for which $\mathsf{H}_{y,x}=0$, there is focusing in the direction normal to the field, conjugate points being diametrically opposite, but there is no focusing in the direction parallel to the field. This is, of course, obvious from simple geometrical considerations. It is likewise obvious, without explicit calculation, that since the radius of curvature of each trajectory is $\mathsf{p}^{-1}\mathsf{H}_y$, an increase $\delta\mathsf{p}$ of beam momentum will displace the image radially outward by a distance $2a(\delta\mathsf{p}/\mathsf{p})$ or, in the usual notation, $2a\mathsf{p}^{-2}\langle 1+\Phi\rangle\epsilon$.

It is possible to obtain identical focusing in the x- and y-directions by prescribing that $\mathsf{H}_{y,x}=\frac{1}{2}\mathsf{p}\kappa^2$. The solutions of the equations (5.4.10) are found to be of the form (5.4.5), where

$$g(z)=\cos(2^{-\frac{1}{2}}\kappa z),\quad h(z)=\sin(2^{-\frac{1}{2}}\kappa z), \tag{5.4.11}$$

and it is now assumed that $z_0=0$, $z_a=\frac{1}{2}\pi a\sqrt{2}$ and $z_b=\pi a\sqrt{2}$ (fig. 5.3). The dispersion may be calculated as in the previous example. We now obtain from (5.2.14), in place of (5.4.2),

$$m^{(1)\mathrm{I}}=-\mathsf{p}^{-1}\kappa\langle 1+\Phi\rangle x_0, \tag{5.4.12}$$

from which we obtain, in place of (5.4.9),

$$x_b^{(0)\mathrm{I}}=-4\mathsf{p}^{-2}\langle 1+\Phi\rangle a,\quad y_b^{(0)\mathrm{I}}=0. \tag{5.4.13}$$

It is seen from the non-relativistic form of (5.4.13) that the dispersion of the stigmatic magnetic spectrograph is the same as that of the stigmatic electrostatic spectrograph; it may also be verified that the astigmatic electrostatic spectrograph which has cylindrical electrodes has the same dispersion as the uniform-field astigmatic magnetic spectrograph.

The above stigmatic magnetic β-ray spectrograph has been built by Svartholm and Siegbahn,‡ who have pointed out that, since

‡ N. Svartholm and K. Siegbahn, *Ark. Mat. Astr. Fys.* **33***a* (1946), no. 21.

$H_{y,x}/H_y = \frac{1}{2}\kappa$, the field strength in the plane of symmetry and in the neighbourhood of the ray axis varies as $r^{-\frac{1}{2}}$, where r measures the radius from the axis of symmetry.

In the spectrographs which we have so far considered, the object and image—i.e. the source and the photographic plate or counter—are situated in the electric or magnetic field. However, it is practically advantageous to have the source and plate situated outside the field; the system then resembles a light-optical spectrometer. The dispersion may be made very high by lengthening the 'arms'

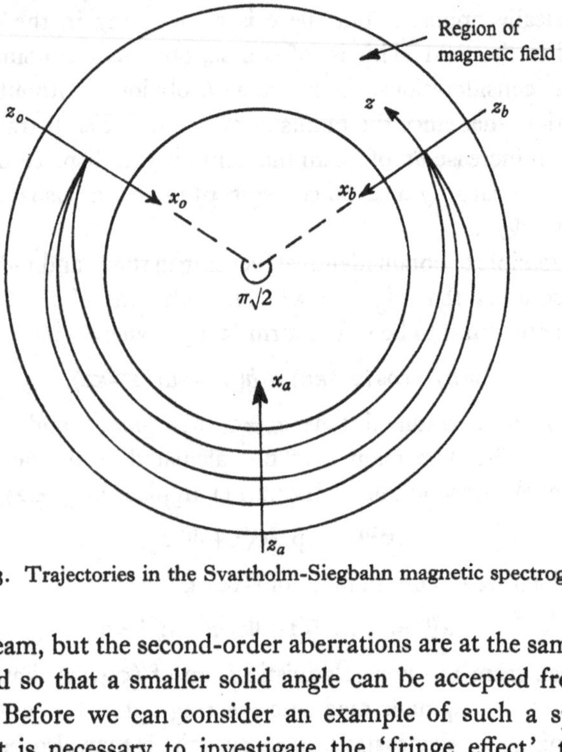

Fig. 5.3. Trajectories in the Svartholm-Siegbahn magnetic spectrograph.

of the beam, but the second-order aberrations are at the same time increased so that a smaller solid angle can be accepted from the source. Before we can consider an example of such a spectrograph, it is necessary to investigate the 'fringe effect', that is, the focusing action of the boundary of the field which will be assumed magnetic.

Suppose an electron beam crosses a boundary at which the magnetic field strength increases suddenly by an amount ΔH_y (fig. 5.4). The ray axis will suffer a discontinuity only in its curvature, but we see from (5.4.10) that neighbouring rays will suffer deflexions of amounts given by

$$\left.\begin{aligned}\Delta x' &= x\mathrm{p}^{-1}\int \mathsf{H}_{y,x}\,\mathrm{d}z,\\ \Delta y' &= -y\mathrm{p}^{-1}\int \mathsf{H}_{y,x}\,\mathrm{d}z,\end{aligned}\right\} \tag{5.4.14}$$

where the range of integration is supposed to extend only over the indefinitely short distance in which the change of field strength takes place. If it is supposed that the beam makes an angle ψ with the normal to the boundary, as shown, the derivative and parameter of integration may be transformed by the relations

$$\left.\begin{aligned}\mathsf{H}_{y,x} &= \mathsf{H}_{y,n}\sin\psi,\\ \mathrm{d}n &= \mathrm{d}z\cos\psi.\end{aligned}\right\} \tag{5.4.15}$$

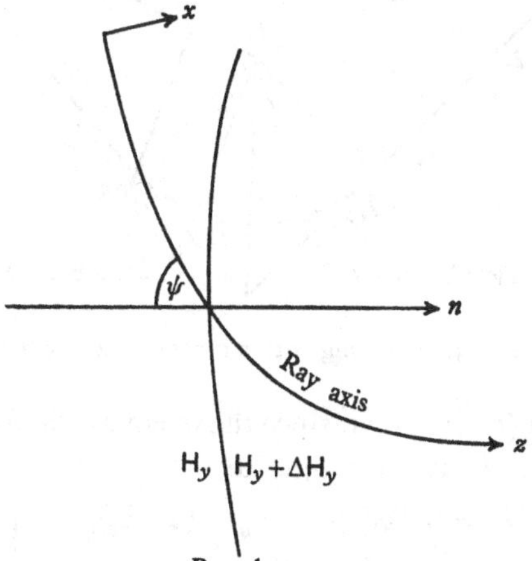

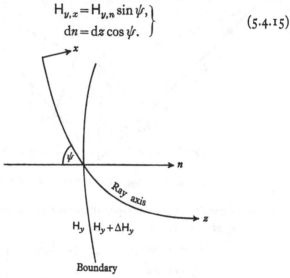

Fig. 5.4. Focusing effect of field discontinuity.

It follows at once from (5.4.14) and (5.4.15) that

$$\left.\begin{aligned}\Delta x' &= p^{-1}\Delta\mathsf{H}_y\tan\psi\,.\,x,\\ \Delta y' &= -p^{-1}\Delta\mathsf{H}_y\tan\psi\,.\,y.\end{aligned}\right\} \tag{5.4.16}$$

Hence there is no focusing action if the beam enters normally to the boundary of the discontinuity, but if ψ is non-zero there is equal and opposite focusing in the x- and y-directions which may be characterized by focal lengths given by

$$\left.\begin{aligned}1/f_x &= -p^{-1}\Delta\mathsf{H}_y\tan\psi,\\ 1/f_y &= p^{-1}\Delta\mathsf{H}_y\tan\psi.\end{aligned}\right\} \tag{5.4.17}$$

Let us now consider the symmetrical magnetic spectrograph shown in fig. 5.5; the angle of deflexion is 2θ and the length of the arms is L. If $R = \mathrm{p}\mathrm{H}_y^{-1}$, the z-co-ordinates of the planes designated in the diagram may be taken to be

$$z_o = -(R\theta + L), \quad z_c = -R\theta, \quad z_a = 0, \quad z_d = R\theta \quad \text{and} \quad z_b = R\theta + L.$$

If the equations of the general paraxial ray are now written as

$$\left.\begin{aligned} x(z) &= x_o g_x(z) + x_a h_x(z), \\ y(z) &= y_o g_y(z) + y_a h_y(z), \end{aligned}\right\} \tag{5.4.18}$$

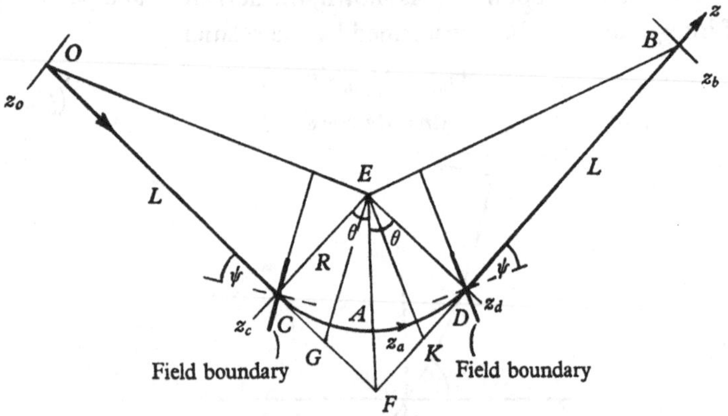

Fig. 5.5. Focusing with a magnetic sector field.

it follows from (5.4.10) that, since the magnetic field is uniform and since $z = z_o$ and $z = z_b$ are conjugate,

$$\left.\begin{aligned} h_x &= L^{-1}\cos\theta(z - z_o) && (z < z_c), \\ h_x &= \cos\kappa z && (z_c < z < z_d), \\ h_x &= L^{-1}\cos\theta(z_b - z) && (z > z_d), \\ h_y &= L^{-1}(z - z_o) && (z < z_c), \\ h_y &= 1 && (z_c < z < z_d), \\ h_y &= L^{-1}(z_b - z) && (z > z_d). \end{aligned}\right\} \tag{5.4.19}$$

However, these rays satisfy (5.4.16) only if

$$L = 2R\cot\theta \quad \text{and} \quad \tan\psi = \tfrac{1}{2}\tan\theta. \tag{5.4.20}$$

It is interesting to note the geometrical interpretation of these relations. If, in fig. 5.5, OC and EA meet in F, and G is the mid-point of CF, then the angle OEG is a right angle and the field boundary at C is parallel to EG.

The dispersion of this spectrograph may be calculated in the usual way. We find from (5.3.5), (5.4.12) and (5.4.18) that

$$V_{ob}^{(1)\mathrm{I}} = -2\mathsf{p}^{-1}\langle 1+\Phi\rangle \sin\theta . x_a. \qquad (5.4.21)$$

Hence, noting that $k^{-1}g_b = -(\mathsf{p}h_b')^{-1}$, we find from (5.3.6) that

$$x_b^{(0)\mathrm{I}} = -4\mathsf{p}^{-2}\langle 1+\Phi\rangle R, \quad y_b^{(0)\mathrm{I}} = 0, \qquad (5.4.22)$$

which shows that the dispersion of this instrument is independent of the angle of deflexion of the beam.

5.5. Mass spectrographs

In this section we are to consider the possibility of designing instruments to discriminate not the energies of particles of known specific charge but the specific charges of particles whose energies may be known only approximately. Such instruments are known as 'mass spectrographs'.

It was pointed out in the introduction to this chapter that if the energy of a beam of particles were known exactly one could determine their masses by measuring their momenta by a deflecting magnetic field such as was considered in the last section. Since, however, energy fluctuations are inherent in any source of ions, more accurate determinations of specific charge may be made with an instrument whose image formation is independent of small changes in beam energy. Mass spectrographs of this type are termed 'double-focusing'.

Mass spectrographs are commonly composed of consecutive electric and magnetic sector fields—similar to the last β-ray spectrograph considered in the preceding section—but it is proposed, for simplicity, that we restrict our attention to a single example in which the fields are superposed.

Since we are no longer considering electrons, it is necessary to reconsider our electric and magnetic units. As was stated in § 1.2, it seems appropriate in dealing with mass spectrographs to suppose that strengths of electric and magnetic fields are measured in e.s.u. and e.m.u., respectively, so that, *in the present context*, ϕ, p *and* **H** *should be regarded as abbreviations for the expressions given in* (1.2.5). It will also be recalled that, since ions have positive charge, and since their masses are large compared with that of an electron, the signs attached to ϕ and **H** will now be changed and relativistic effects will no longer be important.

The formula (5.2.14) for the first-order variational function must therefore be rewritten as

$$m^{(1)} = -(\mathsf{p}\kappa + \mathsf{p}^{-1}\Phi_x + \mathsf{H}_y)\,x \tag{5.5.1}$$

from which we obtain, in place of (5.2.16), the relation

$$\mathsf{p}\kappa + \mathsf{p}^{-1}\Phi_x + \mathsf{H}_y = 0. \tag{5.5.2}$$

It also follows from (5.5.1) that, in the usual notation, the function which determines energy dispersion is given by

$$m^{(1)\mathrm{I}} = (\mathsf{p}^{-1}\kappa - \mathsf{p}^{-3}\Phi_x)\,x. \tag{5.5.3}$$

The function which determines mass dispersion may be found by noting from (1.2.5) and from the relation

$$\mathsf{p} = \sqrt{(-2\Phi)}, \tag{5.5.4}$$

which replaces (1.2.4*a*), the formulae

$$\frac{\partial\Phi}{\partial\eta} = \eta^{-1}\Phi, \quad \frac{\partial\mathsf{p}}{\partial\eta} = -\eta^{-1}\mathsf{p}^{-1}\Phi, \quad \frac{\partial\mathsf{H}_y}{\partial\eta} = \eta^{-1}\mathsf{H}_y, \tag{5.5.5}$$

where $\eta = e/m_0$. The partial derivative of $m^{(1)}$ with respect to η is therefore

$$m^{(1)M} = -\tfrac{1}{2}\eta^{-1}\mathsf{H}_y x, \tag{5.5.6}$$

in deriving which we have used (5.5.2).

It is clear from (5.5.3) that if the condition

$$\Phi_x = -2\kappa\Phi \tag{5.5.7}$$

is imposed, the instrument will be free of zero-order energy dispersion. It is seen from (5.5.6) that the mass dispersion is determined by the magnetic field.

On performing the appropriate modifications in (5.2.15), the paraxial ray equations are found to be

$$\left.\begin{aligned} &\frac{d}{dz}\left(\mathsf{p}\frac{dx}{dz}\right) + (-2\mathsf{p}^{-1}\kappa\Phi_x + \mathsf{p}^{-3}\Phi_x^2 + \mathsf{p}^{-1}\Phi_{xx} - \kappa\mathsf{H}_y + \mathsf{H}_{y,x})\,x = 0,\\ &\frac{d}{dz}\left(\mathsf{p}\frac{dy}{dz}\right) + (-\mathsf{p}^{-1}\Phi'' + \mathsf{p}^{-1}\kappa\Phi_x - \mathsf{p}^{-1}\Phi_{xx} - \mathsf{H}_{y,x})\,y = 0. \end{aligned}\right\} \tag{5.5.8}$$

If we now specify that κ, Φ, Φ_x, etc., should be independent of z and impose the condition (5.5.7), we find—with the help of (5.5.2)

—that the ray equations (5.5.8) reduce to

$$\frac{d^2x}{dz^2}+\kappa^2 x=0, \quad \frac{d^2y}{dz^2}+\kappa^2 y=0, \tag{5.5.9}$$

provided that $$\mathsf{p}^{-1}\Phi_{xx}+\mathsf{H}_{y,x}=0. \tag{5.5.10}$$

Hence if we choose $z_o=0$, $z_a=\frac{1}{2}\pi a$ and $z_b=\pi a$, where $\kappa a=1$, the paraxial rays may be expressed in the form (5.4.5), where $g(z)$ and $h(z)$ are again given by (5.4.6). The instrument is therefore of the 'semi-circular' type.

The mass dispersion may be calculated in the same way as the energy dispersion. We find from (5.3.5), (5.4.5) and (5.5.6) that

$$V_{ob}^{(1)M}=2\eta^{-1}\mathsf{p}x_a, \tag{5.5.11}$$

from which, with the help of (5.3.6), we may deduce that

$$x_b^{(0)M}=2a\eta^{-1}. \tag{5.5.12}$$

This means that if the specific charge of the particles constituting a beam is increased by a certain fraction—i.e. if the mass of the particles is decreased by this fraction—then the image of the entrance slit is displaced inward by the same fraction of the diameter of the instrument.

It is instructive to consider further the chromatic aberration of this particular instrument. Since the function $m^{(1)\mathrm{I}}$ now vanishes identically, the principal chromatic-aberration term in the expansion (5.3.4) is $m^{(2)\mathrm{I}}$, which is responsible for paraxial chromatic aberration. This may now be calculated in the same way as the paraxial chromatic aberration of systems of rotational symmetry, i.e. by a first-order perturbation characteristic function.

We find from (5.2.15) that, for mass spectrographs,

$$m^{(2)\mathrm{I}}=-\tfrac{1}{2}\mathsf{p}^{-1}(x'^2+y'^2)+\tfrac{1}{2}(2\mathsf{p}^{-3}\kappa\Phi_x-3\mathsf{p}^{-5}\Phi_x^2-\mathsf{p}^{-3}\Phi_{xx})x^2 \\ +\tfrac{1}{2}(\mathsf{p}^{-3}\Phi''-\mathsf{p}^{-3}\kappa\Phi_x+\mathsf{p}^{-3}\Phi_{xx})y^2, \tag{5.5.13}$$

which reduces in our case to

$$m^{(2)\mathrm{I}}=-\tfrac{1}{2}\mathsf{p}^{-1}(x'^2+y'^2)+\tfrac{1}{2}(-\mathsf{p}^{-1}\kappa^2-\mathsf{p}^{-3}\Phi_{xx})x^2 \\ +\tfrac{1}{2}(-\mathsf{p}^{-1}\kappa^2+\mathsf{p}^{-3}\Phi_{xx})y^2. \tag{5.5.14}$$

It is seen from this formula that the paraxial chromatic aberration will be stigmatic only if $\Phi_{xx}=0$. If we then note that

$$m^{(2)}=\tfrac{1}{2}\mathsf{p}(x'^2+y'^2)-\tfrac{1}{2}\mathsf{p}\kappa^2(x^2+y^2), \tag{5.5.15}$$

we may make use of the rule established in §§ 2.4 and 3.5 and replace $m^{(2)I}$ by $m^{(2)I} + \mathsf{p}^{-2} m^{(2)}$, i.e. by

$$m^{(2)I} = -\mathsf{p}^{-1}\kappa^2(x^2+y^2). \tag{5.5.16}$$

We now find, in the usual way, that

$$V_{ob}^{(2)I} = -\tfrac{1}{2}\pi\mathsf{p}^{-1}\kappa(x_o^2+y_o^2+x_a^2+y_a^2) \tag{5.5.17}$$

so that

$$x_b^{(1)I} = -\pi\mathsf{p}^{-2}x_a, \quad y_b^{(1)I} = -\pi\mathsf{p}^{-2}y_a. \tag{5.5.18}$$

It follows from (5.5.18) that the only effect of the paraxial chromatic aberration is a defocusing: if the beam energy is increased from $-\Phi$ to $-(1+\varpi)\Phi$, the focus is displaced by a distance $\varpi.\tfrac{1}{2}\pi a$ behind the image plane and a point object will be imaged as a disk of radius $\varpi.\tfrac{1}{2}\pi r_A$, where r_A is the aperture radius. A comparison of this result with (5.5.12) shows that if ions are known to have a fractional spread of energy ϖ, the resolving power of the instrument, expressed as $\delta\eta/\eta$, cannot be better than $\varpi.\tfrac{1}{4}\pi(r_A/a)$

Although it was instructive to consider the choice $\Phi_{xx}=0$, the performance of the spectrograph could be improved by taking $\Phi_{xx}=2\kappa^2\Phi$ for we then have, in place of (5.5.16),

$$m^{(2)I} = -2\mathsf{p}^{-1}\kappa^2 y^2, \tag{5.5.19}$$

so that (5.5.18) is replaced by

$$x_b^{(1)I} = 0, \quad y_b^{(1)I} = -2\pi\mathsf{p}^{-2}y_a. \tag{5.5.20}$$

It follows from these formulae that the image of a point object is now spread into a line parallel to the y-axis. Hence, since the instrument images a slit which is parallel to the y-axis, the resolution will be unaffected by a spread of beam energy as far as we have considered it.

5.6. Cathode-ray-tube deflectors

So far we have considered the optical properties of electron-optical systems in which the beam follows an arbitrary curved path lying in a plane. While this approach was suitable for the study of β-ray spectrographs and mass spectrographs of the type considered in the preceding sections, it is not suitable for the study of electron-optical systems—such as cathode-ray tubes—whose optical properties must be evaluated as functions of the deflexion suffered by the beam. We shall end this chapter with a presentation of the method which is customarily applied to the latter problem. Although the beam is curved, it is not necessary to introduce explicitly a ray axis, so that the method of calculation is more closely related to that of Chapter 4 than that adopted in the earlier sections of the present chapter.

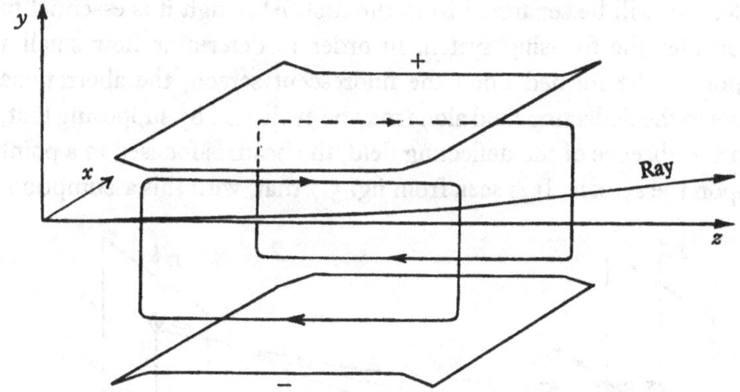

Fig. 5.6. Electrodes and coils producing deflecting field.

It is proposed that we revert to rectangular co-ordinates and adopt the usual convention that the deflected beam lies in the y-z plane. If the electric and magnetic field are symmetrical with respect to both the x-z and y-z planes, they may be conveniently represented by the following potential expansions which satisfy the relevant field equations:

$$\phi = \Phi - \mathsf{E}_0 y - \tfrac{1}{2}\mathsf{E}_2 x^2 y + \tfrac{1}{6}(\mathsf{E}_0'' + \mathsf{E}_2) y^3 + \dots, \qquad (5.6.1)$$

where Φ is now a constant, and

$$\left.\begin{aligned} A_x &= 0 + \dots, \\ A_y &= \tfrac{1}{2}\mathsf{H}_0' x^2 + \dots, \\ A_z &= \mathsf{H}_0 y - \tfrac{1}{2}\mathsf{H}_2 x^2 y + \tfrac{1}{6}\mathsf{H}_2 y^3 + \dots, \end{aligned}\right\} \qquad (5.6.2)$$

for the electric field in the x-z plane and the magnetic field in the y-z plane are then given by

$$E_y(x, 0, z) = \mathsf{E}_0(z) + \tfrac{1}{2}\mathsf{E}_2(z)\, x^2 + \dots \qquad (5.6.3)$$

and

$$H_x(0, y, z) = \mathsf{H}_0(z) + \tfrac{1}{2}\mathsf{H}_2(z)\, y^2 + \dots. \qquad (5.6.4)$$

An electric deflecting field of the type (5.6.1), for which $\phi(x, y, z)$ is odd in y, is known as a 'balanced field'. The coils and electrodes which would produce such fields are indicated in fig. 5.6.

We are assuming for simplicity that the only fields which need be considered are given by (5.6.1) and (5.6.2), but no difficulty is encountered in extending the calculations to take into account the focusing field of an electric or magnetic lens or a second deflecting field which, if magnetic, may be superposed upon the first but, if electric, will be separated from the first. Although it is essential to consider the focusing system in order to determine how small a spot can be formed upon the fluorescent screen, the aberrations due to the deflecting field alone may be evaluated by supposing that, in the absence of the deflecting field, the beam is focused to a point upon the screen. It is seen from fig. 5.7 that, with this assumption,

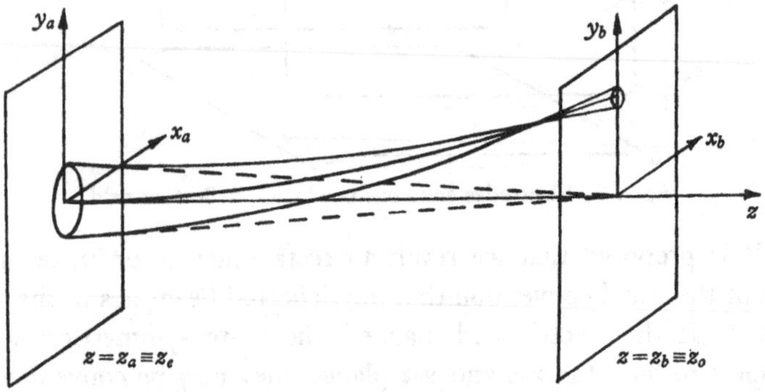

Fig. 5.7. Undeflected and deflected rays.

it is convenient to regard the deflecting field as an electron-optical system which forms an image of a *virtual object*, namely, the point focus of the undeflected beam. In accordance with the notation suggested in § 3.5, m and $\tilde{m}$ will be the variational functions characterizing the deflecting field and free space, respectively; since the aperture must precede the deflecting field, the aperture plane may serve also as the 'entrance plane'.

Since Φ is assumed independent of z, it is convenient to write the formula for the variational function as

$$m = \mathsf{p}^{-1}\{\sqrt{(2\phi)}\sqrt{(1+x'^2+y'^2)} - x'A_x - y'A_y - A_z\}, \quad (5.6.5)$$

where, to the non-relativistic approximation,

$$\mathsf{p} = \sqrt{(2\Phi)}. \quad (5.6.6)$$

By applying (5.6.1) and (5.6.2), we may expand the function in the form (4.3.1), where now

$$m^{(2)} = \tfrac{1}{2}(x'^2+y'^2) - (\mathsf{p}^{-2}\mathsf{E}_0 + \mathsf{p}^{-1}\mathsf{H}_0)y \quad (5.6.7)$$

and

$$m^{(4)} = -\tfrac{1}{8}(x'^2+y'^2)^2 + A(x'^2+y'^2)\,y + Bx^2y' + Cx^2y + Dy^3 + Ey^2, \quad (5.6.8)$$

where

$$\left.\begin{aligned} &A = -\tfrac{1}{2}\mathsf{p}^{-2}\mathsf{E}_0, \quad B = -\tfrac{1}{2}\mathsf{p}^{-1}\mathsf{H}_0', \quad C = -\tfrac{1}{2}(\mathsf{p}^{-2}\mathsf{E}_2 - \mathsf{p}^{-1}\mathsf{H}_2), \\ &D = \tfrac{1}{6}\{\mathsf{p}^{-2}(\mathsf{E}_0'' + \mathsf{E}_2) - \mathsf{p}^{-1}\mathsf{H}_2\}, \quad E = -\tfrac{1}{2}\mathsf{p}^{-4}\mathsf{E}_0^2. \end{aligned}\right\} \quad (5.6.9)$$

In this expansion the functions $\mathsf{E}_0, \mathsf{E}_2, \ldots, \mathsf{H}_0, \mathsf{H}_2$, etc., have been assigned the same order of magnitude as the off-axis co-ordinates x and y. It will be seen that $m^{(2)}$ represents the ideal properties of the deflector so that $m^{(4)}$ must represent the principal aberrations. The functions $\tilde{m}^{(2)}$ and $\tilde{m}^{(4)}$ are obtained from (5.6.7) and (5.6.8) by omitting the field-dependent terms.

Let the aperture plane be $z=z_a$ and the image plane, i.e. the fluorescent screen, $z=z_b$; since the screen is understood to contain the virtual object, $z_o=z_b$ (fig. 5·7). The 'paraxial' ray equations, i.e. the equations derivable from $m^{(2)}$, are

$$\frac{d^2x}{dz^2}=0, \quad \frac{d^2y}{dz^2}+\mathsf{p}^{-2}\mathsf{E}_0+\mathsf{p}^{-1}\mathsf{H}_0=0, \quad (5.6.10)$$

the solution of which may be conveniently written in the form

$$\left.\begin{aligned} x(z) &= x_o g(z) + x_a h(z), \\ y(z) &= y_o g(z) + y_a h(z) + Y(z), \end{aligned}\right\} \quad (5.6.11)$$

where
$$g(z)=\frac{z-z_a}{z_b-z_a}, \quad h(z)=\frac{z_b-z}{z_b-z_a}, \quad (5.6.12)$$

and $Y(z)$ is the particular solution of (5.6.10) for which $y_a=y_a'=0$. $Y'(z)$ and $Y(z)$ may be evaluated by quadrature:

$$Y'(z) = -\int_{z_a}^{z} (\mathsf{p}^{-2}\mathsf{E}_0+\mathsf{p}^{-1}\mathsf{H}_0)\,d\zeta, \quad (5.6.13)$$

and
$$Y(z) = -\int_{z_a}^{z} (\mathsf{p}^{-2}\mathsf{E}_0+\mathsf{p}^{-1}\mathsf{H}_0)(z-\zeta)\,d\zeta. \quad (5.6.14)$$

It is seen from (5.6.11) and (5.6.12) that, since $h_b=0$, the image of the virtual object $x_o=y_o=0$ is given by $x_b=D_x$, $y_b=D_y$, where

$$D_x=0, \quad D_y=Y_b; \quad (5.6.15)$$

since this is a point focus whose displacement from the focus of the undeflected beam is proportional to the strength of the deflecting field, the 'paraxial' approximation yields the ideal properties of the deflector. It is easy to see that the 'virtual' paraxial rays are given by (5.6.11) provided that $Y(z)$ is omitted; hence $\tilde{g}=g$, $\tilde{h}=h$ and $\tilde{Y}=0$.

In evaluating the third-order aberrations of the deflexion system we adopt the formula (3.5.28) which, since $z_e=z_a$ and $z_o=z_b$, may be written as

$$V_{ob}^{(4)}=\int_{z_a}^{z_b} m^{(4)}\,dz-\int_{z_a}^{z_b} \tilde{m}^{(4)}\,dz. \tag{5.6.16}$$

Since the second term on the right-hand side of (5.6.16) cancels the 'field-free' part of the first term, and since the characteristic function must be even in x, the function $V_{ob}^{(4)}(x_a, y_a)$, which characterizes the deflexion aberrations when the undeflected beam has a point focus at $x_o=y_o=0$, may be expressed as

$$V_{ob}^{(4)}=P^*x_a^2y_a+Q^*y_a^3+R^*x_a^2+S^*y_a^2+T^*y_a; \tag{5.6.17}$$

the term independent of x_a and y_a may be ignored.

The aberrations may now be calculated by means of (3.5.21). It is seen from (3.3.36) and (5.6.12) that, since the first term of the formula (5.6.7) is independent of p, $k=(z_b-z_a)^{-1}$ and $g_b=1$, so that

$$\left.\begin{aligned} x_b^{(3)}&=2P^\dagger x_a y_a+2R^\dagger x_a,\\ y_b^{(3)}&=P^\dagger x_a^2+3Q^\dagger y_a^2+2S^\dagger y_a+T^\dagger,\end{aligned}\right\} \tag{5.6.18}$$

where

$$P^\dagger=(z_b-z_a)P^*, \quad \text{etc.} \tag{5.6.19}$$

Since the expressions (5.6.17) is of the fourth order in the off-axis co-ordinates and the strength of the deflecting field, it must be possible to rewrite (5.6.18) in the form

$$\left.\begin{aligned} x_b^{(3)}&=2PD_y x_a y_a+2RD_y^2 x_a,\\ y_b^{(3)}&=PD_y x_a^2+3QD_y y_a^2+2SD_y^2 y_a+TD_y^3,\end{aligned}\right\} \tag{5.6.20}$$

where P, Q, etc., which are related to $P^\dagger$, $Q^\dagger$, etc., by

$$\left.\begin{aligned} P=P^\dagger/D_y,\quad Q=Q^\dagger/D_y,\quad R=R^\dagger/D_y^2,\\ S=S^\dagger/D_y^2,\quad T=T^\dagger/D_y^3,\end{aligned}\right\} \tag{5.6.21}$$

are independent of the magnitude of the deflexion.

The geometrical significance of the coefficients P, Q, etc., may be established by the method used in §4.3. It is found that P and Q together give rise to a type of *coma* whose pattern depends on the relative values of P and Q; its magnitude varies linearly with the deflexion and quadratically with the aperture. The coefficients R and S represent both *astigmatism* and *field curvature*; the astigmatic difference is $2(z_b - z_a) \mid R-S \mid D_y^2$, and the radius of curvature of the screen necessary for best definition is $\frac{1}{2}(z_b - z_a)^{-1} \mid R+S \mid^{-1}$; the screen is convex or concave as seen from the deflector according as $R+S$ is positive or negative. The remaining coefficient T represents a pure *distortion* in the sense that if T is non-zero the deflexion does not vary linearly with the strength of the deflecting field. The relative importance of the aberrations is determined, to a great extent, by the power of D_y by which the coefficients are multiplied in (5.6.20), for the maximum deflexion is usually much larger than the beam aperture. Hence distortion, astigmatism and field curvature are the most important aberrations; coma may usually be neglected.

The coefficients of the deflexion aberrations may be calculated by substituting the formulae (5.6.11), with $x_o = y_o = 0$, in (5.6.8) and integrating, the field-independent terms being ignored in accordance with (5.6.16). Hence we obtain the formulae

$$\left.\begin{aligned}
P^* &= \int_{z_a}^{z_b} \{-\tfrac{1}{2}h'^3 Y' + Ah'^2 h + Bh'h^2 + Ch^3\}\, dz,\\
Q^* &= \int_{z_a}^{z_b} \{-\tfrac{1}{2}h'^3 Y' + Ah'^2 h + Dh^3\}\, dz,\\
R^* &= \int_{z_a}^{z_b} \{-\tfrac{1}{4}h'^2 Y'^2 + Ah'^2 Y + Bh^2 Y' + Ch^2 Y\}\, dz,\\
S^* &= \int_{z_a}^{z_b} \{-\tfrac{3}{4}h'^2 Y'^2 + A(h'^2 Y + 2h'hY') + 3Dh^2 Y + Eh^2\}\, dz,\\
T^* &= \int_{z_a}^{z_b} \{-\tfrac{1}{2}h' Y'^3 + A(2h'Y'Y + hY'^2) + 3DhY^2 + 2EhY\}\, dz.
\end{aligned}\right\} \quad (5.6.22)$$

It is interesting to note from (5.6.9) and (5.6.22) that since $C+3D$ is independent of E_2 and H_2, the combinations $P+3Q$ and $R+S$, the latter of which represents field curvature, are completely determined by the 'paraxial' field functions $\mathsf{E}_0(z)$ and $\mathsf{H}_0(z)$. The significance of this result will be seen later. We may also note that whereas, if the field is purely electric, the three terms A, D and E contribute to $P+3Q$ and $R+S$, the single term B contributes to these combinations when the field is purely magnetic; this suggests what is, in fact, an accepted rule, that *of two comparable deflectors, one electric and one magnetic, the latter has the lower aberrations.*

Let us consider as an example the deflexion of a beam by a balanced electric field. We shall suppose for simplicity that the field is sharply defined in the x-z plane, so that

$$\mathsf{E}_0(z)=\mathsf{E} \quad \text{for} \quad 0<z<l, \qquad \mathsf{E}_0(z)=0 \quad \text{for} \quad z<0,\ z>l. \tag{5.6.23}$$

In writing $z_a=0$, $z_b=L$, we imply that both aperture and image planes are in field-free space, the former being adjacent to the field boundary.

The second of equations (5.6.10) is readily integrated to give

$$\left.\begin{aligned} Y(z) &= -\tfrac{1}{2}\mathsf{p}^{-2}\mathsf{E}z^2 \quad \text{for} \quad z<l, \\ &= -\mathsf{p}^{-2}\mathsf{E}l(z-\tfrac{1}{2}l) \quad \text{for} \quad z>l, \end{aligned}\right\} \tag{5.6.24}$$

so that, with the help of (5.6.6), we obtain

$$D_y = -\tfrac{1}{2}(\mathsf{E}/\Phi)\, l(L-\tfrac{1}{2}l). \tag{5.6.25}$$

It is seen from the above formulae that, to the paraxial approximation, the beam appears to have been deflected by an angle $-\tfrac{1}{2}(\mathsf{E}/\Phi)\, l$ at the centre of the deflecting field.

It follows from (5.6.3) that if the boundary of the field in the x-z plane is assumed to be straight where it intersects the z-axis, $\mathsf{E}_2(z)\equiv 0$. The coefficients of the deflexion aberrations may now be evaluated by means of (5.6.9), (5.6.22), (5.6.19) and (5.6.21). One may take account of the term E_0'' in (5.6.9), which will be infinite at the 'fringe' of the field, either by performing integrations by parts or, equivalently, by adopting Dirac's delta notation‡ and noting that

$$\mathsf{E}_0'(z)=\mathsf{E}\{\delta(z)-\delta(z-l)\}. \tag{5.6.26}$$

In this way we find that

$$\left.\begin{aligned} &P=L^{-2}, \qquad Q=0, \qquad R=-\tfrac{1}{4}L^{-2}(1-\lambda)(1-\tfrac{1}{2}\lambda)^{-2}, \\ &S=-L^{-2}\lambda^{-1}(1-\tfrac{3}{4}\lambda+\tfrac{1}{4}\lambda^2)(1-\tfrac{1}{2}\lambda)^{-2}, \quad T=\tfrac{1}{8}L^{-2}\lambda(1-\tfrac{1}{2}\lambda)^{-3}, \end{aligned}\right\} \tag{5.6.27}$$

where we have written $\lambda=l/L$ for brevity. We find from (5.6.27) that the astigmatic difference is approximately $2D_y^2/l$ and that, for best definition, the screen should be concave, as seen from the

‡ P. A. M. Dirac, *Quantum Mechanics* (Oxford University Press, 3rd ed. 1947), p. 58.

deflector, with a radius of curvature of approximately $\frac{1}{2}l$. It is obvious that if, as is usually the case, the deflexion is comparable with the length of the deflecting field, the aberrations arising from the terms R and S will be very important. For example, if $L=25$ cm., $l=5$ cm., $D_y=10$ cm. and the aperture is of radius 0·1 cm., the spot becomes an ellipse whose length in the y-direction is 0·32 cm. but whose width in the x-direction is only 0·016 cm. The displacement of the centre of this spot from its 'ideal' position is only 0·04 cm., and coma is quite unimportant for the dimensions of the coma pattern would be about 0·0003 cm.

It is interesting to consider the effect of shaping the deflector electrodes. If the plates are so shaped that lines of constant field strength drawn in the x-z plane have a radius of curvature ρ where they cross the z-axis, then $\mathsf{E}_2=\mathsf{E}_0'/\rho$ or $-\mathsf{E}_0'/\rho$, according as the lines are convex or concave as seen from the image screen. By applying this rule to the foregoing example, we find that the distortion may be eliminated by so shaping the field boundary at $z=l$ that the centre of the osculating circle coincides with the centre of the image screen. However, it would probably be preferable to eliminate the astigmatism, and this may be achieved by making the centre of the osculating circle lie approximately at the centre of the deflecting field; the field curvature is, of course, unchanged, but the distortion coefficient T is now approximately $\frac{1}{4}L^{-2}$. Hence, for the values adopted above, the spot becomes a circular disk of diameter 0·16 cm. whose centre is displaced by 0·4 cm. from its 'ideal' position. It is possible that if the image screen were curved and the aperture were very large, coma might become important; it is therefore worth remarking that P and Q are the only coefficients which vary with the curvature of the field boundary at $z=0$ and that the coma pattern is smallest when $P=3Q$. It is found that the coma is minimized in the above example by arranging that the boundary at $z=0$ has approximately the same curvature as the boundary at $z=l$.

PART II. DYNAMIC ELECTRON OPTICS

CHAPTER 6

UNIFORM FOCUSING IN PARTICLE ACCELERATORS

6.1. Introduction

The second part of this monograph, on 'dynamic' electron optics, should deal, in full generality, with the focusing of beams of electrons, or other charged particles, by any electromagnetic fields which are varying in time, but it is proposed to restrict our attention to the specific problem of focusing in particle accelerators. It might be supposed that since it will be necessary to introduce time explicitly, all connexion with optics must be lost, but this is not true. Even if one were to adopt the standard viewpoint which accords to time a preferential role (that of a parameter which enumerates a sequence of kinematical configurations of a dynamical system), it would be found that the concepts of geometrical optics are repeated in classical dynamics for the simple reason that Fermat's principle and Hamilton's principle are mathematically identical in form. The dynamical discoveries for which Sir William Hamilton is remembered were indeed foreshadowed in his theory of geometrical optics. There is, however, no obligation to give time preference over the other dynamical variables, and we shall find that it is advantageous to extend to the treatment of particle accelerators the methods which were developed in the earlier chapters of this monograph for application to static electron optics; with this viewpoint, one regards time as 'just another co-ordinate'.

The introduction of one more co-ordinate does not mark the *essential* difference between the two parts of this monograph, for we shall find, in Chapter 7, that one may carry out many important calculations concerning particle accelerators without introducing time at all; such calculations might, on this account, be classed as 'static electron optics'. The significant distinction is to be found in the respective interpretations of the term 'focusing'. In static electron optics 'focusing' implies, although it is not synonymous with, 'image formation'; in the study of particle accelerators 'focusing' is much more closely related to the dynamical concept of 'stability'. One may to some extent characterize the difference by stating that 'focusing' in static electron optics and in the theory of particle accelerators relates to the collective behaviour of trajectories in beams of finite and of infinite lengths respectively.

It is instructive to return now to the assumptions, set out in §1.1, which underlie the theory of geometrical electron optics. One of these, the requirement that the field should be static, is no longer relevant, but the remainder will still be adopted. Consequences of the wave nature of electrons do not arise, so that classical mechanics is perfectly appropriate. The long-range Coulomb and Lorentz forces may be important if a heavy beam is injected at low energy, in which case one may employ the method of §4.5, but they are never important once the particles are under way. The short-range effects of statistical fluctuations of the field due to 'granulation' of the beam are normally unimportant but, on the other hand, one must bear in mind the possibility of gas scattering. Radiation reaction is negligible in the acceleration of heavy particles but becomes important in the acceleration of electrons to high energies in circular accelerators;‡ for this very reason the highest accelerations of electrons are effected in linear accelerators where radiation reaction may again be neglected.

If one is prepared to accept the restrictions indicated, it is possible to investigate particle accelerators by considering the motion of a single electron in the electromagnetic field of the machine. This motion is determined by the appropriate form of Hamilton's principle which, as has already been indicated, is a variational equation which is usually written in a form analogous to (3.2.1) but which could equally well be written in the form (1.3.9).

The division of material into two chapters has been dictated by the recent development in the design of accelerators which goes under various names but which was originally termed 'the strong-focusing principle' by its proponents.§ The present chapter will deal with particle accelerators in which the acceleration and focusing are 'uniform' in the sense that their character is essentially the same at all times although their strengths will normally vary slowly over the acceleration cycle.

‡ See refs. (8), (45) and (48).

§ See ref. (55).

It will be seen that focusing in particle accelerators may be analysed by a procedure similar to that adopted in Chapters 3 and 5 in the study of electron-optical systems with curved axes. One may adopt some trajectory as a 'trajectory axis' and so set up co-ordinates which are appropriate for the study of small clusters of electrons, for a beam of space-time trajectories appears in three-dimensional section as a cluster of particles. There will of course be three 'off-axis' co-ordinates where before there were only two; two of the three are associated with the spatial motion, the other with the 'bunching' or 'phase' motion. By carrying out expansions in these off-axis co-ordinates, one is able to obtain a linear approximation to the motion which is good for small enough clusters; the higher non-linear terms may be treated by perturbation methods similar to those used earlier in the study of aberrations. Non-linear terms introduce coupling between the modes of 'paraxial' oscillation and also a dependency of the frequency of oscillation on amplitude; it is important to verify, in designing a machine, that these effects do not result in uncontrolled expansion of the beam. A study of the effect of constructional errors is important for the same reason, but such calculations, which are analogous to those of §4.6, may also be carried out by perturbation methods.

It was stated earlier that in this chapter the acceleration and focusing will be assumed to change only slowly over the acceleration cycle. This assumption is acceptable in the study of circular accelerators, for in these a particle usually experiences some hundreds of phase oscillations and some tens of thousands of transverse oscillations during their acceleration, but is not so appropriate to linear accelerators, for in these a particle may execute only a few oscillations during acceleration. Now there are certain quantities, which are closely related to the Poincaré invariant introduced in §2.3, whose values are unaltered if the relevant dynamical system undergoes a slow aperiodic perturbation; these are known as 'adiabatic invariants', and it will be found that they offer a powerful method of analysing the behaviour of particle accelerators. Instead of tracing trajectories over the complete acceleration cycle, one may establish the modes of oscillation for a typical section of the accelerator, which may be taken to be so short that parameters such as the mean energy of the beam of particles do not change appreciably over its length, and then by constructing the adiabatic invariants one may determine how the amplitudes of these modes of oscillation vary during acceleration.

Two examples will be discussed in this chapter: the synchrotron, which is a circular accelerator, and the linear electron accelerator. The analysis of the properties of these machines by the methods of this chapter is of course based on the assumption that the behaviour of the machines would not be significantly affected if the electromagnetic fields (which are in fact produced by localized resonant cavities, wave-guide cavities and coils) were smoothed out so as not to vary appreciably over short distances. This assumption is not too unrealistic for the machines indicated, but one would nevertheless feel safer if the theory were capable of taking account of the discrete structure of the fields. Such a theory is to be developed in the following chapter.

6.2. The variational equation

This section will be devoted to the first step in the analysis of the focusing properties of a particle accelerator, namely, the setting up of the appropriate variational equation and the separation of the integrand into a series of functions which describe the linear (paraxial) behaviour and the non-linear effects. The appropriate variational equation is now Hamilton's principle, which is usually written in the form

$$\delta \int_A^B L\,dt = 0, \tag{6.2.1}$$

where L is the Lagrangian function which, for an electron moving in an electromagnetic field, is expressed as $L(\mathbf{x}, \dot{\mathbf{x}}, t)$ and is given by (1.3.2), i.e. by

$$L = 1 - \sqrt{(1 - \dot{\mathbf{x}}^2)} + \phi - \dot{\mathbf{x}}.\mathbf{A}. \tag{6.2.2}$$

It will be remembered that in writing the Lagrangian in the above form we have adopted electron-optical units, whose values were established in § 1.2, for the electric and magnetic field strengths, and we have also chosen that unit of time which makes the velocity of light unity so that *time is measured in centimetres*; with this unit of time, the period of an electromagnetic wave is numerically equal to its free-space wavelength. The reader may also be reminded that the values of 'proton' units were given at the end of § 1.2, and that the signs attached to ϕ and $\mathbf{A}$ in (6.2.1) must be reversed if one is considering particles with positive charge. The terminal conditions implicit in (6.2.1) are familiar and are analogous to those of (1.3.4); the end-points of the path which is undergoing variation are fixed in space-time.

If we now subject (6.2.1) to the same operations as (3.2.1) in §3.2, we are led to introduce a vector canonically conjugate to the co-ordinate vector $\mathbf{x}$ defined by

$$\mathbf{p}=\frac{\partial L}{\partial \dot{\mathbf{x}}}. \tag{6.2.3}$$

This is the *canonical momentum vector* which we find from (6.2.2) to be expressible as

$$\mathbf{p}=(1+e)\,\dot{\mathbf{x}}-\mathbf{A}, \tag{6.2.4}$$

where it is now necessary to introduce the symbol e for the kinetic energy of the particle, so that

$$e=(1-v^2)^{-\frac{1}{2}}-1, \tag{6.2.5}$$

since in a time-varying electromagnetic field e may no longer be identified with the electric potential ϕ. The scalar symbol p will still denote the scalar kinetic momentum which was introduced in §1.2, but the formula (1.2.4) must now be replaced by

$$p=\sqrt{(2e+e^2)}. \tag{6.2.6}$$

It is easy to prove, from (6.2.5) and (6.2.6), that (6.2.4) may be rewritten as (2.2.8) and so establish the identity of the 'ray vector' and the 'canonical momentum vector'.

In place of the ray equation (3.2.6) we now have the familiar equation of motion

$$\frac{\mathrm{d}}{\mathrm{d}t}\left(\frac{\partial L}{\partial \dot{\mathbf{x}}}\right)=\frac{\partial L}{\partial \mathbf{x}}. \tag{6.2.7}$$

A *characteristic function* $V(\mathbf{x}_a, t_a, \mathbf{x}_b, t_b)$ may be defined as

$$V=\int_A^B L\,\mathrm{d}t, \tag{6.2.8}$$

the integral now being taken along the trajectory connecting the designated points of space-time. If we now allow the spatial positions of the terminal points to vary but keep their times fixed, we obtain once more the differential relation (3.2.7) except that the summation must now be extended over the three space co-ordinates. However, it is interesting to allow the terminal times also to vary; on noting that this additional variation adds $L\delta t$ to δV but also displaces the trajectory in time so that the variation at the original time is no longer $\delta\mathbf{x}$ but $\delta\mathbf{x}-\dot{\mathbf{x}}\,\delta t$, we see that the more general form of *Hamilton's differential relation* is

$$\delta V=(\mathbf{p}_b\,.\,\delta\mathbf{x}_b+p_{tb}\,\delta t_b)-(\mathbf{p}_a\,.\,\delta\mathbf{x}_a+p_{ta}\,\delta t_a) \tag{6.2.9}$$

where p_t, which is clearly the variable canonically conjugate to the time, is given by

$$p_t=L-\mathbf{p}\,.\,\dot{\mathbf{x}}. \tag{6.2.10}$$

The expression on the right-hand side of (6.2.10) will be recognized as the negative of the total energy of the particle so that, since in our units ϕ is the negative of the potential energy, we may write

$$p_t=-e+\phi. \tag{6.2.11}$$

This may be verified directly from (6.2.2). We have chosen to use the symbol e for the kinetic rather than the total energy, since it is the former which is of direct physical interest. We have chosen to represent the total energy by the symbol $-p_t$, since the only interest which it will have for us is due to the fact that its negative is the variable canonically conjugate to the time.

Although it is customary to adopt time as the independent variable of the dynamical equations, this choice is obligatory only if the particles are stationary somewhere along their paths as, for instance, in electron mirrors. In undertaking an investigation of particle accelerators, one is well advised to consider as an alternative choice for the independent variable the co-ordinate which measures distance along the beam. (The difference is the same as that between taking photographs of a racecourse at regular intervals of time and placing photo-finish cameras along its length.) If the electromagnetic field varies as simply in space as it does in time, neither method holds a great advantage over the other; but if its variation along the beam is much more complicated than its variation in time (which is normally sinusoidal), and this is the case in many 'conventional' accelerators and all strong-focusing accelerators, it is simpler to make the second choice, since the functions which represent the complicated variation then appear naturally as functions of the independent variable. It is clear that even if the field of the accelerator varies simply along the beam, the second technique is better suited to consideration of the injection and extraction mechanisms.

For the above reasons we shall proceed to set up a system of curvilinear co-ordinates, based upon a 'trajectory axis', and to adopt spatial distance along this trajectory as the independent variable. The trajectory which it is proposed to take as basis of the co-ordinate system determines a curve in space, and we may base a set of spatial co-ordinates on this curve exactly as in §3.2 by adopting this curve as the z-axis and setting up planes normal to this axis which we map by rectangular co-ordinates x, y or, alternatively, x_i where $i=1,2$ (see fig. 3.1). However, the trajectory is a curve in space-time, as indicated in fig. 6.1, so that we may conveniently relabel the time co-ordinate so that its origin always lies

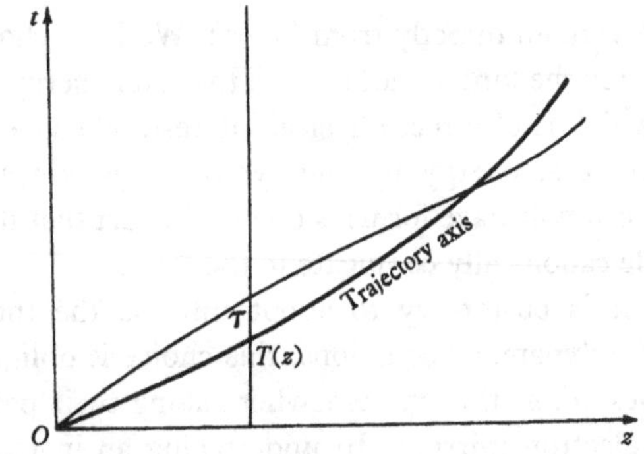

Fig. 6.1. Projection of the trajectory axis and an arbitrary trajectory upon the z-t plane.

on the trajectory axis. If τ is the new time variable, then

$$t = T(z) + \tau \tag{6.2.12}$$

if $t = T(z)$ upon the trajectory axis; in terms of the co-ordinates x, y, τ, z, the trajectory axis has the equation $x = y = \tau = 0$.

The variational equation (6.2.1) may now be rewritten as

$$\delta \int_A^B M \, \mathrm{d}z = 0 \tag{6.2.13}$$

where M, which we shall refer to as the *variational function*, is related to L by

$$M = L \frac{\mathrm{d}t}{\mathrm{d}z} \tag{6.2.14}$$

but is expressed as $M(x_i, \tau, x_i', \tau', z)$, where a prime has its usual significance, differentiation with respect to z. The Euler-Lagrange equations derivable from M are

$$\frac{\mathrm{d}}{\mathrm{d}z}\left(\frac{\partial M}{\partial x_i'}\right)=\frac{\partial M}{\partial x_i},\quad \frac{\mathrm{d}}{\mathrm{d}z}\left(\frac{\partial M}{\partial \tau'}\right)=\frac{\partial M}{\partial \tau},\tag{6.2.15}$$

which will be referred to as the *trajectory equations.*

If the spatial set (x, y, z) are rectangular co-ordinates, as they will be in a linear accelerator, then it follows from (6.2.2) that, ignoring the constant term,

$$M=-\sqrt{[(T'+\tau')^2-x_i'^2-1]}+(T'+\tau')\,\phi-x_i'A_i-A_z.\tag{6.2.16}$$

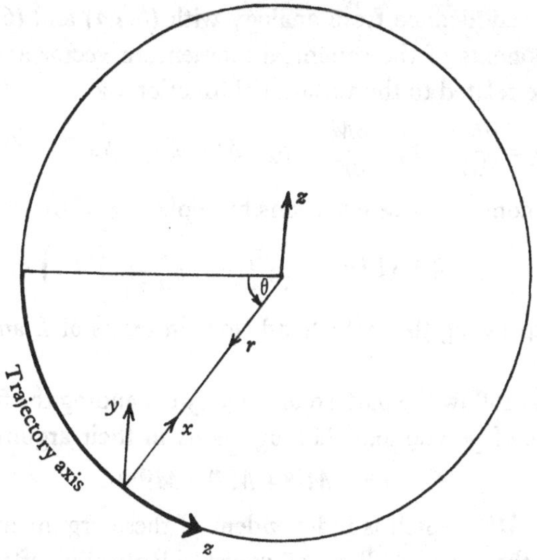

Fig. 6.2. Alternative co-ordinates for the description of a circular system.

However, if one were considering a circular accelerator or a 'race track', the co-ordinates would no longer be rectangular although they might still be chosen to be orthogonal. One could make the choice of § 5.2, in which case (6.2.16) would be replaced by

$$M=-\sqrt{[(T'+\tau')^2-x'^2-y'^2-(1-\kappa x)^2]}$$
$$+(T'+\tau')\,\phi-x'A_x-y'A_y-(1-\kappa x)\,A_z.\tag{6.2.17}$$

This may be re-expressed in terms of the more familiar cylindrical polar co-ordinates, should the machine be circular, by replacing (x, y, z) by $(a-r, z, a\theta)$ and (A_x, A_y, A_z) by $(-A_r, A_z, A_\theta)$, where $a=\kappa^{-1}$ (cf. fig. 6.2).

It is clear from (6.2.14) that we can form the same characteristic function V as was defined by (6.2.8) by the equation

$$V=\int_A^B M\,\mathrm{d}z. \tag{6.2.18}$$

The differential relation (6.2.9) must still hold good, but its form will be changed slightly because of (6.2.12); we now have

$$\delta V=\{p_{ib}\delta x_{ib}+(p_{zb}+p_{\tau b}T_b')\,\delta z_b+p_{\tau b}\delta\tau_b\} \\ -\{p_{ia}\delta x_{ia}+(p_{za}+p_{\tau a}T_a')\,\delta z_a+p_{\tau a}\delta\tau_a\}, \tag{6.2.19}$$

where p_τ is identical with p_t, the negative of the total energy. Moreover, we may deduce from analogy with (6.2.3) and (6.2.10) that the components of the canonical momentum vector and the total energy are related to the variational function by

$$p_i=\frac{\partial M}{\partial x_i'},\quad p_\tau=\frac{\partial M}{\partial \tau'},\quad p_z=M-p_i x_i'-p_\tau(T'+\tau'). \tag{6.2.20}$$

One may confirm these equations by replacing M by

$$(T'+\tau')\,L\left(x_i, z, \frac{x_i'}{T'+\tau'}, \frac{1}{T'+\tau'}, T+\tau\right)$$

and so expressing the right-hand sides in terms of L and its derivatives.

We may follow the pattern of § 3.3 by expanding the function M as a series of polynomials homogeneous in their arguments x_i, τ, x_i', τ':

$$M=M^{(0)}+M^{(1)}+M^{(2)}+M^{(3)}+\dots. \tag{6.2.21}$$

The term $M^{(0)}$, which is independent of these arguments, may be ignored; the term $M^{(1)}$ is to vanish identically, after possible application of the partial-integration rule of § 3.2, as the condition that the trajectory axis should be a possible trajectory; the term $M^{(2)}$ therefore determines the paraxial behaviour, i.e. the limiting properties of the trajectories as the beam width becomes indefinitely small. As before, the quadratic terms of the variational function yield the linear approximation to the equations of motion; the higher terms, $M^{(3)}$, etc., represent the non-linear corrections to the linear approximation.

As in § 3.3, it is possible to establish the coefficients of the terms of $M^{(2)}$ which are quadratic in x_i', τ'; these terms are derived from the first term of (6.2.17). If we note that, upon the trajectory axis,

$$v=1/T', \tag{6.2.22}$$

we find that the scalar kinetic momentum and kinetic energy upon the trajectory axis are given by

$$\mathsf{p}=(T'^2-1)^{-\frac{1}{2}} \tag{6.2.23}$$

and

$$\mathsf{e}=(T'^2-1)^{-\frac{1}{2}}T'-1, \tag{6.2.24}$$

so that

$$T'=\mathsf{p}^{-1}(1+\mathsf{e}) \tag{6.2.25}$$

and

$$M^{(2)}=\tfrac{1}{2}\mathsf{p}x_i'^2+\tfrac{1}{2}\mathsf{p}^3\tau'^2+\text{other terms}; \tag{6.2.26}$$

terms not given explicitly in (6.2.26) involve x_i and τ.

Now it is clear from (6.2.11) and (6.2.20) that the kinetic energy e is the negative of the partial derivative with respect to τ' of the kinetic part of M, that is, of the first term of (6.2.17). The paraxial contribution to the kinetic energy is readily found to be

$$e^{(1)}=-\mathsf{p}^3\tau'-\mathsf{p}^2(1+\mathsf{e})\kappa x. \tag{6.2.27}$$

It is an immediate consequence of (6.2.11) that the paraxial part of p_τ is expressible as

$$p_\tau^{(1)}=-e^{(1)}+\phi^{(1)}, \tag{6.2.28}$$

where $\phi^{(1)}$ denotes the paraxial contribution to ϕ; hence if there is no static electric field, so that it is possible to eliminate the scalar potential, $p_\tau^{(1)}$ is simply the negative of $e^{(1)}$.

Now the function $M^{(2)}$ may appear as the sum of three terms, each of which involves only one of the co-ordinates x_1, x_2 and τ; in this case it is 'orthogonal' in the sense of § 3.3 and may be expressed in a form similar to (3.3.4):

$$M^{(2)}=\tfrac{1}{2}(\mathsf{p}x_1'^2-\mathsf{X}_1x_1^2)+\tfrac{1}{2}(\mathsf{p}x_2'^2-\mathsf{X}_2x_2^2)+\tfrac{1}{2}(\mathsf{p}^3\tau'^2-\mathsf{T}\tau^2). \tag{6.2.29}$$

It is clear that, since we are supposing that the coefficients in (6.2.29) vary slowly, the beam will be 'focused' only if X_1, X_2 and T are all positive, for the three trajectory equations

$$\frac{\mathrm{d}}{\mathrm{d}z}\left(\mathsf{p}\frac{\mathrm{d}x_1}{\mathrm{d}z}\right)+\mathsf{X}_1x_1=0,\quad \frac{\mathrm{d}}{\mathrm{d}z}\left(\mathsf{p}\frac{\mathrm{d}x_2}{\mathrm{d}z}\right)+\mathsf{X}_2x_2=0,\quad \frac{\mathrm{d}}{\mathrm{d}z}\left(\mathsf{p}^3\frac{\mathrm{d}\tau}{\mathrm{d}z}\right)+\mathsf{T}\tau=0, \tag{6.2.30}$$

derivable from the variational function (6.2.29) then have bounded solutions, expressible in terms of circular functions, instead of unbounded solutions expressible in terms of hyperbolic functions.

Let us now suppose that there is coupling between two of the three co-ordinates, say between x_1 and x_2. The general form of that part of the variational function which refers to these two co-ordinates may be written as‡

$$M = \tfrac{1}{2}\begin{pmatrix} x_i' \\ x_i \end{pmatrix}' \begin{pmatrix} a_{ij} & c_{ij} \\ c_{ji} & b_{ij} \end{pmatrix} \begin{pmatrix} x_j' \\ x_j \end{pmatrix}, \qquad (6.2.31)$$

where a_{ij} and b_{ij} are symmetrical. We may derive the appropriate trajectory equations from (6.2.31) by means of (6.2.15), and it is proposed that we seek 'eigen solutions', possibly complex, of the form

$$x_i(z) = X_i e^{\lambda z}. \qquad (6.2.32)$$

We find that the amplitudes X_i must satisfy the homogeneous linear equations

$$[a_{ij}\lambda^2 + (c_{ij} - c_{ji})\lambda - b_{ij}]\, X_j = 0 \qquad (6.2.33)$$

so that the *characteristic exponents*, i.e. the possible values of λ, are the roots of the *characteristic equation*

$$\| a_{ij}\lambda^2 + (c_{ij} - c_{ji})\lambda - b_{ij} \| = 0. \qquad (6.2.34)$$

If the roots of this equation are all different, the general solution of the trajectory equations is the general linear combination of the eigen solutions, and it follows that if these eigen solutions are all bounded, every solution is bounded. Hence we may say that *the system is focused if the characteristic exponents are all purely imaginary and if no two are equal.*

If two of the characteristic exponents are equal, further investigation is called for. It may be that the system is still focused; this would obviously be true of the trivial case for which the matrix in (6.2.31) is diagonal. But it may also be that the general solution is unbounded; for instance, if one of the trajectory equations is $x'' = 0$, the root $\lambda = 0$ occurs twice but, although the corresponding eigen solutions are bounded, the general solution of the trajectory equation is $x(z) = az + b$, which is unbounded.

On remembering that the transpose of a matrix has the same determinant as the matrix itself, we see from (6.2.34) that *if λ is a root, so is $-\lambda$*, so that (6.2.34) must be a quadratic equation in λ^2. Explicitly, the equation is

$$(a_{11}a_{22} - a_{12}^2)\lambda^4 - [a_{11}b_{22} + a_{22}b_{11} - 2a_{12}b_{12} - (c_{12} - c_{21})^2]\lambda^2 - (b_{11}b_{22} - b_{12}^2) = 0, \qquad (6.2.35)$$

‡ It will be remembered that a prime denotes a transpose.

and a sufficient condition that the system be focused is that the roots of this equation for λ^2 should be real, negative and different.

The above theory may be applied to the case of coupling between all three co-ordinates merely by extending the range of summation.

6.3. The adiabatic invariants

It has already been stated in the introduction to this chapter that the analysis of the properties of particle accelerators is greatly facilitated by the use of 'adiabatic invariants'. These are functions of the variables used to describe the oscillatory process which are unchanged by a sufficiently slow aperiodic perturbation. We have seen in the preceding section that the motion of a particle in a focused particle accelerator is an oscillatory process; the acceleration itself, which entails a slow variation of such quantities as the mass of the particles and the strength of the deflecting field, may be regarded as 'adiabatic' perturbation. The evaluation of the appropriate adiabatic invariants may therefore be expected to show how the amplitudes of the oscillations vary during the course of the acceleration.

Adiabatic invariants are closely related to Poincaré invariants which have already been discussed in §2.3. One may, indeed, regard the adiabatic approximation as the approximation entailed in identifying one invariant with the other; this approach leads to the interpretation of the adiabatic approximation as the assumption that there are no 'preferred' phases of the oscillatory process under discussion. It will be shown that the 'phase-independence' hypothesis is equivalent to the usual conditions for the existence of adiabatic invariants.

For simplicity, we begin this analysis with the more familiar choice of time as the independent variable. We shall restrict our attention to a particle with co-ordinates x_r moving in a field of force, but our treatment may be applied to a more general dynamical system simply by replacing x_r by generalized co-ordinates. We assume that the field does not depend explicitly upon the time but varies with some parameter (or parameters) c. It will be convenient, for this particular section, to introduce the Hamiltonian rather than the Lagrangian function; the appropriate Hamiltonian may be written as

$$\mathscr{H} = \mathscr{H}(x_r, p_s; c). \qquad (6.3.1)$$

The well-known canonical equations are

$$\frac{\mathrm{d}x_r}{\mathrm{d}t}=\frac{\partial \mathscr{H}}{\partial p_r}, \quad \frac{\mathrm{d}p_r}{\mathrm{d}t}=-\frac{\partial \mathscr{H}}{\partial x_r}. \tag{6.3.2}$$

We suppose that the general solution of these equations of motion is periodic or multiply periodic with periods $2\pi/\omega_i(c)$, so that it may be expressed as

$$\left.\begin{aligned} x_r&=x_r(\omega_i(c)\,t+\kappa_i, K_k, t;\, c),\\ p_r&=p_r(\omega_i(c)\,t+\kappa_i, K_k, t;\, c),\end{aligned}\right\} \tag{6.3.3}$$

where each of the functions on the right-hand side is periodic in its first argument; K_k denotes the parameters which are required in addition to the phases κ_i to enumerate the trajectories.‡ Clearly the range of values of i together with the range of values of k must be twice the number of degrees of freedom of the system, that is, for a particle, six.

Now suppose that c is not a constant but is a slowly varying function of time $c(t)$. We may still retain the expression (6.3.3) for the equation of the general trajectory but the parameters κ_i and K_k, which denote a particular trajectory, must now be supposed to vary slowly with time so that if at and near a time t_a the parameters have values κ_{ia}, K_{ka}, at and near a distant time t_b the parameters will have values κ_{ib}, K_{kb}. The general functional relation may, of course, be written as

$$\left.\begin{aligned} \kappa_{ib}&=\kappa_{ia}+f_i(\kappa_{ja}, K_{ha};\, t_a, t_b),\\ K_{kb}&=g_k(\kappa_{ja}, K_{ha};\, t_a, t_b);\end{aligned}\right\} \tag{6.3.4}$$

we note that f_i and g_k must be cyclic in the phases κ_{ja}.

Now since the expression (6.3.3) is cyclic in each of the phases κ_i, we may form a tube of trajectories by allowing κ_{ia} to run over the range 0 to 2π, and we may form another tube of trajectories by allowing κ_{ib} to run over the same range; although they may always have one trajectory in common, these tubes will generally not coincide (see fig. 6.3). We saw in § 2.3 that any tube of rays in a static electron-optical system has associated with it a Poincaré invariant which is formed by taking the integral $\oint p_r\,\mathrm{d}x_r$ over any path em-

‡ The following example may help to clarify the present usage of the term 'multiply-periodic' and satisfy the reader who feels that the argument t in (6.3.3) is redundant: $x_1=K_1 \sin(\omega t+\kappa_1)$, $x_2=K_2+K_3 t$; p_1, p_2 accordingly.

bracing the tube. We may deduce from a comparison of (6.2.1) with (3.2.1) that in any dynamical system a tube of trajectories has associated with it a Poincaré invariant which is formed by taking the integral $\oint (p_r \mathrm{d}x_r + p_t \mathrm{d}t)$ over any path embracing the tube. Hence the tubes which we have formed have invariant integrals P_{ia} and P_{ib}, where

$$\left.\begin{aligned} P_{ia} &= \oint \mathrm{d}\kappa_{ia} \left(p_r \frac{\partial x_r}{\partial \kappa_{ia}} + p_t \frac{\partial t}{\partial \kappa_{ia}} \right), \\ P_{ib} &= \oint \mathrm{d}\kappa_{ib} \left(p_r \frac{\partial x_r}{\partial \kappa_{ib}} + p_t \frac{\partial t}{\partial \kappa_{ib}} \right) \end{aligned}\right\} \quad (i \text{ n.t.b.s.}). \qquad (6.3.5)$$

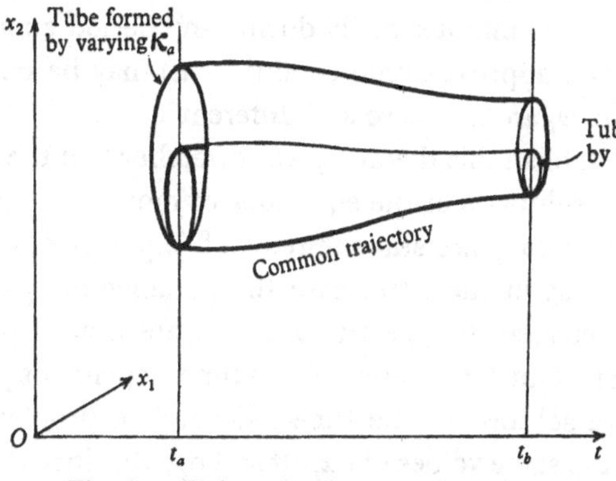

Fig. 6.3. Tubes of trajectories formed by varying a phase parameter at different times.

(It will be remembered that 'n.t.b.s.' is an abbreviation for 'not to be summed'.) We shall prove in this section that *under certain conditions these two tubes coincide* so that, in consequence, the two invariants of (6.3.5) may be identified as one invariant, A_i, which will then be termed an *adiabatic invariant*:

$$\mathsf{A}_i = \oint \mathrm{d}\kappa_i \left(p_r \frac{\partial x_r}{\partial \kappa_i} + p_t \frac{\partial t}{\partial \kappa_i} \right) \quad (i \text{ n.t.b.s.}). \qquad (6.3.6)$$

We shall first show that a sufficient condition for the existence of adiabatic invariants is that the functional form of the relations (6.3.4) should be preserved if the zero points from which the phases κ_i are measured suffer arbitrary displacements; this was referred to earlier as the 'phase-independence' hypothesis. This hypothesis leads to the following more restricted form for the relations (6.3.4):

$$\kappa_{ib} = \kappa_{ia} + \phi_i(K_{ha}), \quad K_{kb} = K_{kb}(K_{ha}). \tag{6.3.7}$$

Hence each κ_{ib} is simply a displacement of the corresponding κ_{ia}, where the displacement may depend upon the parameters K_{ka}. It follows immediately that κ_{ia} and κ_{ib} enumerate trajectories of the same tube so that the expression (6.3.6) is an invariant.

It remains to demonstrate that the validity of the above hypothesis concerning the phases is implied by the following conditions:

(*a*) the system is not degenerate, i.e. no linear combination of the ω_i with integral coefficients vanishes;

(*b*) $c(t)$ is aperiodic; and

(*c*) $c(t)$ varies only infinitesimally during any period $2\pi\omega_i^{-1}$.

To the *paraxial* approximation condition (*a*) may be relaxed to read simply 'the ω_i are non-zero and different'.

In order to establish this theorem, we remember that the expressions (6.3.3) are solutions of the equations of motion (6.3.2) if c is constant, and that they are still solutions of (6.3.2) if c is varying with time but that in the latter case the parameters κ_i and K_k, which will be referred to collectively as α_m, are time-dependent. Hence if, at an instant t, a solution of the time-dependent problem coincides with a solution of the time-independent problem, both solutions have the same values of dx_r/dt and dp_r/dt, since the right-hand sides of the equations (6.3.2) will take the same values in each case. But in one case the parameters α_m and c are varying, whereas in the other they are fixed; we deduce that, at the intersection of the solutions to the two problems, the contributions to dx_r/dt and dp_r/dt due to the variation of the parameters α_m and c must vanish, i.e.

$$\left.\begin{aligned} \frac{\partial p_r}{\partial \alpha_m}\dot{\alpha}_m + \frac{\partial p_r}{\partial c}\dot{c} = 0, \\ \frac{\partial x_r}{\partial \alpha_m}\dot{\alpha}_m + \frac{\partial x_r}{\partial c}\dot{c} = 0. \end{aligned}\right\} \tag{6.3.8}$$

These six equations for $\dot{\alpha}_m$ may be reformed as

$$\dot{\alpha}_m\{\alpha_m, \alpha_n\} = \{\alpha_n, c\}\,\dot{c}, \tag{6.3.9}$$

where we have used the notation (2.3.3) for the Lagrange bracket. The equations (6.3.9) may now be interpreted as a reformulation of the equations of motion for the time-dependent problem in terms of the parameters α_m.

If c is a constant, an interpretation of §2.3 for dynamical rather than electron-optical systems shows that the Lagrange brackets $\{\alpha_m, \alpha_n\}$ are constants; we may infer that if c varies slowly $\{\alpha_m, \alpha_n\}$ will vary only slowly. If the parameters α_m enumerate the trajectories completely and uniquely, it must be possible to invert the matrix $\{\alpha_m, \alpha_n\}$ so as to obtain from (6.3.9), upon integration,

$$\alpha_{mb} - \alpha_{ma} = \int_{t_a}^{t_b} dt\{\alpha_n, c\}\,\dot{c}\{\alpha_n, \alpha_m\}^{-1}. \tag{6.3.10}$$

Moreover, the elements of the inverse matrix must be slowly varying so that, since $c(t)$ is aperiodic, the only oscillatory contribution to the integrand is contained in $\{\alpha_n, c\}$. This term may be expanded as a multiple Fourier series in the cyclic arguments $\omega_i(c)\,t + \kappa_i$ but, since the integral is supposed to extend over a large number of periods and since we are assuming that the system is not degenerate, the only important contribution will come from the term which is *independent* of these arguments. Hence the integrals (6.3.10) are independent of the *phases* characterizing the trajectory. We may deduce at once that the functions f_i and g_k of (6.3.4) are independent of the phases, so that our original hypothesis to this effect is justified provided that the relevant conditions (*a*), (*b*) and (*c*) are satisfied. To the *paraxial* approximation the functions (6.3.3) are linear in their periodic arguments so that $\{\alpha_n, c\}$ is only quadratic in the circular functions with arguments $\omega_i(c)\,t + \kappa_i$. The simplification of condition (*a*) appropriate to this approximation follows immediately.

For our purpose, the most convenient form of the formula (6.3.6) is obtained by adopting as contour of integration the section of the tube of trajectories by a plane $t = \text{const}$. Thus

$$\mathsf{A}_i = \oint d\kappa_i p_r \frac{\partial x_r}{\partial \kappa_i} \quad (i \text{ n.t.b.s.}). \tag{6.3.11}$$

However, it is interesting to see that the formula can be put into the familiar form for the special case of a system which is only singly periodic. For this case we choose as contour of integration not the curve γ_1 (see fig. 6.4) but the contour formed by the curves γ_2 and γ_3, where γ_2 follows a trajectory over one cycle and γ_3 is the straight line joining the ends of this trajectory. Since the system we are now considering is time-independent, the tube of trajectories formed by varying the phase parameter may also be formed by displacing a trajectory in time; the tube is therefore a cylinder with generators parallel to the t-axis so that γ_3 lies in the tube as required. Now

$$\mathsf{A}=\int_{\gamma_3}(p_r\,\mathrm{d}x_r+p_t\,\mathrm{d}t)+\int_{\gamma_3}(p_r\,\mathrm{d}x_r+p_t\,\mathrm{d}t), \qquad (6.3.12)$$

but since $x_r=$ const. along γ_3, the first term of the second integral vanishes. But, again since the Hamiltonian is time-independent,

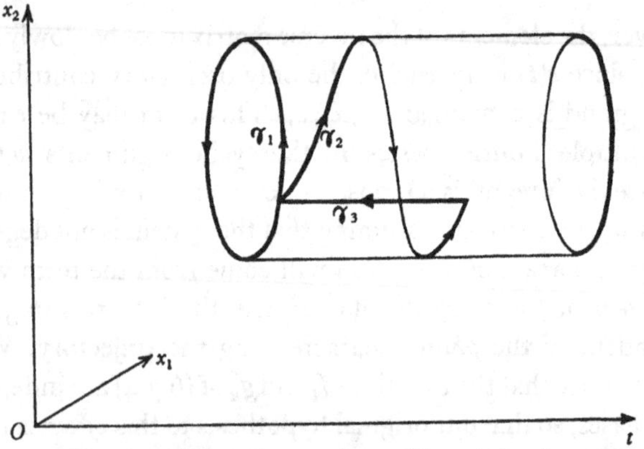

Fig. 6.4. Alternative contours of integration for evaluation of the adiabatic invariant.

the total energy $-p_t$ is constant for any trajectory; hence p_t is constant over the entire tube. The second terms in the two integrals therefore cancel and we are left with

$$\mathsf{A}=\int_{\gamma_3}p_r\dot{x}_r\,\mathrm{d}t; \qquad (6.3.13)$$

this is the integral over one cycle in time of twice the kinetic energy of the system, which is a familiar adiabatic invariant.‡

There is no difficulty in returning to the co-ordinates which were introduced in §6.2. The independent variable is now z and the dependent variables are x_i and τ, with corresponding canonical momenta p_z, p_i and p_τ. The periodicities will now be regarded as periodicities in z rather than t, so that the cyclic arguments appearing

‡ See, for instance, J. H. Jeans, *The Dynamical Theory of Gases* (Cambridge University Press, 4th ed. 1925), p. 415.

in (6.3.3) would now be $\omega_l(c)\,z+\kappa_l$; the 'wavelengths' associated with these periodicities are clearly $2\pi\omega_l^{-1}$. The most useful formula for the adiabatic invariant will be that corresponding to (6.3.11); this is

$$\mathsf{A}_l=\oint \mathrm{d}\kappa_l\left(p_i\frac{\partial x_i}{\partial\kappa_l}+p_\tau\frac{\partial\tau}{\partial\kappa_l}\right)\quad (l \text{ n.t.b.s.}),\qquad (6.3.14)$$

where it is now convenient to use l to enumerate the periodicities.

Consider, as a simple example, the variational function given by the first term of (6.2.29) which we write as

$$M=\tfrac{1}{2}\mathsf{p}(x'^2+\omega^2x^2).\qquad (6.3.15)$$

If p and ω are both constants, the general trajectory may be written as

$$x(z)=K\sin(\omega z+\kappa),\qquad (6.3.16)$$

so that the variation of the corresponding momentum is given by

$$p_x(z)=\mathsf{p}\omega K\cos(\omega z+\kappa).\qquad (6.3.17)$$

The adiabatic invariant is therefore

$$\mathsf{A}\equiv\oint \mathrm{d}\kappa\, p_x\frac{\partial x}{\partial\kappa}=\pi\mathsf{p}\omega K^2.\qquad (6.3.18)$$

Hence if p and ω vary slowly and aperiodically, the amplitude of the trajectory will vary as $(\mathsf{p}\omega)^{-\frac{1}{2}}$. This well-known result may of course be obtained more directly by other methods;‡ the advantage of the method which we have adopted is that it is applicable to systems which are multiply-periodic and that, as we shall see in the next chapter, it may be taken over to the study of dynamical systems which involve fields varying periodically along the trajectories.

An interesting and important problem is that of determining the *acceptance* of an accelerator, i.e. of determining the precise volume of phase space in which the points representative of the particles of a beam must lie in order that they should traverse the whole accelerator. This would enable one to decide what radius and angular divergence of the beam is permissible at injection, or to estimate what fraction of an injected beam could be expected to emerge from the accelerator if its initial radius and angular divergence are large. This problem may be answered simply if one adopts the paraxial approximation and if the variational function is then of the orthogonal form (6.2.29), for one may then consider each off-axis co-ordinate separately and so make use of the results of the preceding paragraph.

‡ See, for instance, H. and B. S. Jeffrey's, *Methods of Mathematical Physics* (Cambridge University Press, 1946), p. 490.

Let us suppose that the geometry of the accelerator entails that a particle is lost if $|x|$ exceeds a value a. Then we see from (6.3.16) and (6.3.18) that, for accepted particles,

$$\mathsf{A} \leqslant \pi a^2 \min(\mathsf{p}\omega). \tag{6.3.19}$$

The accepted area of p_x-x space is now determined by the right-hand side of (6.3.19). The shape of this accepted area at injection is found from (6.3.16) and (6.3.17) to be the ellipse whose equation, expressed in terms of x_o and x'_o, is

$$x_o^2 + \omega_o^{-2} x_o'^2 = \pi a^2 \min(\mathsf{p}\omega/\mathsf{p}_0\omega_o). \tag{6.3.20}$$

One may estimate in a similar manner the accepted area of p_τ-τ space at injection. If the injected beam is steady, we know that its representation in phase space is independent of τ; we also see from (6.2.28) that, in the absence of electrostatic fields, p_τ measures the kinetic energy of the beam.

6.4. The synchrotron

It is proposed that, as an example of the application of the theory which has been set out in the preceding sections of this chapter, we should make an elementary investigation of the synchrotron. This accelerator comprises two essential parts: a circular magnet, similar to that of the β-ray spectrograph of Siegbahn and Svartholm discussed in § 5.4, which guides a beam of electrons along a circular path; and one or more electrode structures, fed from a suitable source, which produces a periodic longitudinal electric field whose purpose is to accelerate the particles. Since the momentum and speed of the particles vary during their acceleration, it is necessary to vary the strength of the magnetic field in order to retain the beam in the correct 'orbit' and to vary the period of the accelerating field in order to keep it 'in step' with the particles. We shall see that, apart from the usual focusing effect of the magnetic field, there is a 'longitudinal' or 'phase' focusing which ensures that certain groups of electrons will remain 'bunched'. By using the adiabatic invariants introduced in the last section, we shall be able to see how the dimensions of these groups change during acceleration.

We shall adopt the curvilinear co-ordinates (x, y, z) suggested in § 6.2. Since the trajectory axis will be taken to be a circle of radius a, these co-ordinates are related by cylindrical polar co-ordinates as shown in fig. 6.2 and the curvature $\kappa = a^{-1}$. We saw in § 1.5 that a magnetic field of rotational symmetry is derivable from a vector potential with only one component A_θ; the same result may be seen from (5.2.9), for if all the coefficients of the vector potential are independent of z, only those coefficients which appear in the expansion of A_z survive. We find from (5.2.6) and (5.2.9) that

$$A_z = -\mathsf{H}x - \tfrac{1}{2}(1+n)\,a^{-1}\mathsf{H}x^2 + \tfrac{1}{2}na^{-1}\mathsf{H}y^2 - \dots \qquad (6.4.1)$$

if we write the values of H_y and $H_{y,x}$ on the trajectory axis as H and $na^{-1}\mathsf{H}$ in accordance with the usual convention, and if we ignore the value of A_z upon the trajectory axis. The last condition is of no account if the field is static, but if, as in a particle accelerator, the magnetic field is changing in time, to neglect the value of the vector potential upon the trajectory axis is to make an approximation since a changing magnetic field produces an electric field which can bring about 'induction' or 'betatron' acceleration. In the present example we are assuming that betatron acceleration is negligible compared with the acceleration due to the imposed accelerating field, but we could take account of the betatron acceleration by replacing (6.4.1) by

$$A_z = \mathsf{A}_z + (a^{-1}\mathsf{A}_z - \mathsf{H})\,x + \tfrac{1}{2}(2a^{-2}\mathsf{A}_z - (1+n)\,a^{-1}\mathsf{H})\,x^2 + \tfrac{1}{2}na^{-1}\mathsf{H}y^2 + \dots, \qquad (6.4.1\,a)$$

where $2\pi a\mathsf{A}_z$ is the magnetic flux embraced by the trajectory axis.

Since the field equations set out in § 1.2 are appropriate only for static fields, it is necessary to write out the field equations for time-varying fields before considering the periodic accelerating electric field. With our choice of units, Maxwell's equations take the form

$$\left.\begin{aligned} \operatorname{div}\mathbf{E} &= \rho, & \operatorname{curl}\mathbf{E} &= -\partial\mathbf{H}/\partial t, \\ \operatorname{div}\mathbf{H} &= 0, & \operatorname{curl}\mathbf{H} &= \mathbf{j} + \partial\mathbf{E}/\partial t. \end{aligned}\right\} \qquad (6.4.2)$$

We may therefore introduce scalar and vector potentials such that

$$\mathbf{E} = -\operatorname{grad}\phi - \partial\mathbf{A}/\partial t, \quad \mathbf{H} = \operatorname{curl}\mathbf{A}, \qquad (6.4.3)$$

and, provided that we specify that

$$\partial\phi/\partial t + \operatorname{div}\mathbf{A} = 0, \qquad (6.4.4)$$

the field equations reduce to

$$\left.\begin{aligned} \nabla^2\phi - \partial^2\phi/\partial t^2 &= -\rho, \\ \nabla^2\mathbf{A} - \partial^2\mathbf{A}/\partial t^2 &= -\mathbf{j}. \end{aligned}\right\} \qquad (6.4.5)$$

If we neglect the effect of space charge and space current, we may put $\phi \equiv 0$ and so characterize the field by a vector potential only which satisfies

$$\operatorname{div} \mathbf{A} = 0 \tag{6.4.6}$$

and

$$\nabla^2 \mathbf{A} - \partial^2 \mathbf{A}/\partial t^2 = 0, \tag{6.4.7}$$

and to which the field vectors are related by

$$\mathbf{E} = -\partial \mathbf{A}/\partial t, \quad \mathbf{H} = \operatorname{curl} \mathbf{A}. \tag{6.4.8}$$

If the free-space wavelength of the periodic electric field is $2\pi\lambda\!\!\!^{-}$, the time-dependence of this field is that of a circular function with argument $\lambda\!\!\!^{-\,-1}t$. The field may also be analysed as a Fourier series in the co-ordinate θ, i.e. $a^{-1}z$, but it is proposed that the field be smoothed out by neglecting all but the first term of this series, that which has period 2π. By an appropriate correlation of the zeros of t and z, the value of the electric field strength on the trajectory axis may be written as

$$E_z = -\mathsf{E} \cos (\lambda\!\!\!^{-\,-1}t - a^{-1}z). \tag{6.4.9}$$

We see from (6.4.8) that the vector potential upon the trajectory axis must be given by

$$A_z = \lambda\!\!\!^{-}\mathsf{E} \sin (\lambda\!\!\!^{-\,-1}t - a^{-1}z). \tag{6.4.10}$$

It would be possible to introduce the other components of the vector potential, which are essential if we are to satisfy (6.4.6), and to expand all three components as power series in x and y but this is not profitable since the influence of the extra terms upon the focusing of the beam is negligible compared with the focusing effect of the magnetic field; the term already given in (6.4.10) is the only one whose effect is not overshadowed by that of the magnetic field. It is, however, essential to expand (6.4.10) in powers of τ; we obtain

$$A_z = \lambda\!\!\!^{-}\mathsf{E} \sin \phi + \mathsf{E} \cos \phi \,.\, \tau - \tfrac{1}{2}\lambda\!\!\!^{-\,-1}\mathsf{E} \sin \phi \,.\, \tau^2 + \ldots \tag{6.4.11}$$

if we assume that, upon the trajectory axis,‡

$$\lambda\!\!\!^{-\,-1}T - a^{-1}z \equiv \phi, \tag{6.4.12}$$

which implies that the particle which describes the reference trajectory rides at a point on the travelling wave which is $\phi/2\pi$ of a wavelength behind the crest. It is strictly necessary to carry out a similar expansion of the part of the vector potential which describes the magnetic field, and this would be essential if one took

‡ ϕ, which henceforth denotes phase, should not be confused with ϕ which has hitherto denoted electric potential.

account of the betatron acceleration, but we shall now assume that such effects are negligible and so leave (6.4.1) in its present form.

Since (6.4.12) holds for all z, $T' = \lambda\!\!\!^{-}/a$ so that, from (6.2.25),

$$\lambda\!\!\!^{-} = \mathsf{p}^{-1}(1+\mathsf{e})\,a. \qquad (6.4.13)$$

We may also derive from (6.4.12) an expression for the phase $\phi^{(1)}$ by which an electron lags behind that describing the trajectory axis:

$$\phi^{(1)} = \lambda\!\!\!^{-\,-1}\tau. \qquad (6.4.14)$$

We may now proceed with the dynamical investigation of the synchrotron by inserting (6.4.1) and (6.4.11) in (6.2.17) and expanding the variational function in terms of the off-axis co-ordinates. We find that

$$M^{(1)} = -(1+\mathsf{e})\,\tau' - \mathsf{E}\cos\phi\,.\,\tau + (\mathsf{H} - \mathsf{p}a^{-1})\,x \qquad (6.4.15)$$

(where we have again neglected a term involving E in comparison with a term involving H), and so derive, from the condition that $x=y=\tau=0$ should be a trajectory, the relation

$$\mathsf{H} = a^{-1}\mathsf{p} \qquad (6.4.16)$$

and the formula

$$\frac{d\mathsf{e}}{dz} = \mathsf{E}\cos\phi, \qquad (6.4.17)$$

which shows that the beam acquires an energy increment $2\pi a\mathsf{E}\cos\phi$ per revolution, as we should expect from (6.4.9) and (6.4.12).

The paraxial contribution to M is given by

$$M^{(2)} = \tfrac{1}{2}\mathsf{p}x'^2 + \tfrac{1}{2}\mathsf{p}y'^2 + \tfrac{1}{2}\mathsf{p}^3\tau'^2 + \mathsf{p}^2(1+\mathsf{e})\,a^{-1}\tau' x + \tfrac{1}{2}\mathsf{p}(n+\mathsf{p}^2)\,a^{-2}x^2$$
$$- \tfrac{1}{2}n\mathsf{p}a^{-2}y^2 + \mathsf{E}a^{-1}\cos\phi\,.\,\tau x + \tfrac{1}{2}\mathsf{p}(1+\mathsf{e})^{-1}a^{-1}\mathsf{E}\sin\phi\,.\,\tau^2. \qquad (6.4.18)$$

We see at once that the paraxial trajectory equation in y may be written in the form

$$\frac{d^2y}{d\theta^2} + ny = 0, \qquad (6.4.19)$$

so that if n is positive there is focusing in the y-direction.

We see from (6.4.18) that there is coupling between the x and τ co-ordinates so that it is necessary to adopt the method set out at the end of § 6.2 for analysing the focusing in these directions. The appropriate part of (6.4.18) may be expressed in the form (6.2.31) with the following values for the matrix elements:

$$\left.\begin{aligned} &a_{xx} = \mathsf{p}, \quad b_{xx} = \mathsf{p}(n+\mathsf{p}^2)\,a^{-2}, \\ &\qquad\quad b_{x\tau} = b_{\tau x} = \mathsf{E}a^{-1}\cos\phi, \qquad c_{\tau x} = \mathsf{p}^2(1+\mathsf{e})\,a^{-1}; \\ &a_{\tau\tau} = \mathsf{p}^3, \quad b_{\tau\tau} = \mathsf{p}(1+\mathsf{e})^{-1}a^{-1}\mathsf{E}\sin\phi, \end{aligned}\right\} \qquad (6.4.20)$$

all other elements vanish. If we insert these values in (6.2.35) and regard E as small, we obtain to first approximation the values λ_B and λ_S for the characteristic exponents where

$$\lambda_B^2 = -(1-n)\,a^{-2}, \quad \lambda_S^2 = -\frac{(n+\mathsf{p}^2)\,\mathsf{E}\sin\phi}{(1-n)\,\mathsf{p}^2(1+\mathsf{e})\,a}. \qquad (6.4.21)$$

In order to have focusing in both the x- and τ-directions, both λ_B and λ_S must be imaginary, so that we must have $n<1$ and $\sin\phi>0$. Hence $0<n<1$ and, if we are to have acceleration, $0<\phi<\frac{1}{2}\pi$.

If we write

$$\lambda = \mathrm{i}a^{-1}\Omega, \qquad (6.4.22)$$

the principal modes will have periodicities of Ω_B^{-1} and Ω_S^{-1} revolutions; with the help of (6.4.17), we see that

$$\Omega_B^2 = 1-n, \quad \Omega_S^2 = \frac{(n+\mathsf{p}^2)\tan\phi}{(1-n)\,\mathsf{p}^2(1+\mathsf{e})}\frac{d\mathsf{e}}{d\theta}. \qquad (6.4.23)$$

The first of these modes will be recognized, from the first of these formulae, as representing the same radial oscillations as one would have in a static magnetic field without an accelerating electric field (cf. § 5.4); these are known as the 'betatron oscillations', since they are the only radial oscillations which arise in the betatron. However, the second mode also appears if there is a travelling electric wave; it is therefore known as the 'synchrotron oscillation'; we see from the second of the above formulae that if the particles execute N revolutions during their acceleration cycle, the period of the synchrotron oscillations is, for high energies, of the order of $\sqrt{N}$ revolutions.

In order to determine the forms of the principal modes, we must solve the equations (6.2.33) for the ratio of‡ T to X, inserting the values of λ given in (6.4.21). If $\lambda=\lambda_B$, we find that

$$T/X = \mathrm{i}\mathsf{p}^{-1}(1+\mathsf{e})\,\Omega_B^{-1} \qquad (6.4.24)$$

from which, with the help of (6.2.32) and (6.4.22), we find that the general form of the betatron oscillations may be written as

$$\left.\begin{aligned} x_B &= K_B\cos(\Omega_B\theta+\kappa_B),\\ \tau_B &= -\mathsf{p}^{-1}(1+\mathsf{e})\,\Omega_B^{-1}K_B\sin(\Omega_B\theta+\kappa_B).\end{aligned}\right\} \qquad (6.4.25)$$

‡ T, as it appears in this context, should not be confused with T as it is used in (6.2.12).

It would appear from (6.2.27) and (6.4.25) that the kinetic energy does not vary as a particle executes betatron oscillations, but this result cannot be expected to hold accurately since we have neglected the higher terms in x and y in the expansion (6.4.11). The magnitude of the phase variation associated with the betatron oscillation may be found with the help of (6.4.13) and (6.4.14) and may be written as

$$|\phi_B^{(1)}| = \Omega_B^{-1} a^{-1} |x_B|. \qquad (6.4.26)$$

Since with normal constructions the variation of x must be small compared with a, the betatron phase oscillation must be small.

We find in the same way for the synchrotron oscillations that

$$T/X = \mathrm{i}\mathsf{p}^{-1}(n+\mathsf{p}^2)(1+\mathsf{e})^{-1}\Omega_S^{-1}, \qquad (6.4.27)$$

so that the general form of the synchrotron oscillation is

$$\left.\begin{aligned} x_S &= K_S \cos(\Omega_S \theta + \kappa_S), \\ \tau_S &= -\mathsf{p}^{-1}(n+\mathsf{p}^2)(1+\mathsf{e})^{-1}\Omega_S^{-1} K_S \sin(\Omega_S\theta + \kappa_S), \end{aligned}\right\} \qquad (6.4.28)$$

and the variation in phase is now given by

$$|\phi_S^{(1)}| = \frac{n+\mathsf{p}^2}{1+\mathsf{p}^2}\Omega_S^{-1} a^{-1} |x_S|. \qquad (6.4.29)$$

Since Ω_S is small, $|\phi_S^{(1)}|$ may be large even though $|x_S|$ must be small compared with a. The energy variation is now non-zero; its magnitude is found from (6.2.27) to be given by

$$|e_S^{(1)}| = (1-n)\,\mathsf{p}^2(1+\mathsf{e})^{-1} a^{-1} |x_S|. \qquad (6.4.30)$$

Hence the fractional variation in energy due to the synchrotron oscillations is of the same order of magnitude as the fractional variation in radius.

There is no difficulty in evaluating the adiabatic invariants associated with the various modes of oscillation. The variation of the y-component of the betatron oscillations may indeed be deduced from the example given at the end of § 6.3, for the terms of (6.4.18) involving y are of the form (6.3.15), where $\omega = n^{\frac{1}{2}} a^{-1}$; hence

$$|y| \propto \mathsf{p}^{-\frac{1}{2}}. \qquad (6.4.31)$$

In order to evaluate the other adiabatic invariants, it is necessary to introduce the paraxial approximations to the momentum components p_x and p_τ. We find from (6.2.20) and (6.4.18) that

$$p_x = \mathsf{p}x', \quad p_\tau = \mathsf{p}^3\tau' + \mathsf{p}^2(1+\mathsf{e})\,a^{-1}x. \qquad (6.4.32)$$

(The second formula might have been taken from (6.2.27) and (6.2.28), since in this problem the electric scalar potential vanishes.) If we now apply (6.3.14) to (6.4.25), we obtain

$$\mathsf{A}_B = -\pi \mathsf{p}\Omega_B a^{-1} K_B^2 \tag{6.4.33}$$

as an adiabatic invariant, so that

$$|x_B| \propto \mathsf{p}^{-\frac{1}{2}}. \tag{6.4.34}$$

If the same formula is now applied to (6.4.28), we obtain

$$\mathsf{A}_S = -\pi(1-n)\,\mathsf{p}(n+\mathsf{p}^2)\,(1+\mathsf{p}^2)^{-1}\Omega_S^{-1}a^{-1}K_S^2. \tag{6.4.35}$$

By using (6.4.23), (6.4.29) and (6.4.30) we find that

$$\left.\begin{aligned} |x_S| &\propto \mathsf{p}^{-1}(n+\mathsf{p}^2)^{-\frac{1}{4}}(1+\mathsf{e})^{\frac{3}{4}},\\ |\phi_S^{(1)}| &\propto (n+\mathsf{p}^2)^{\frac{1}{4}}(1+\mathsf{e})^{-\frac{3}{4}},\\ |e_S^{(1)}| &\propto \mathsf{p}(n+\mathsf{p}^2)^{-\frac{1}{4}}(1+\mathsf{e})^{-\frac{1}{4}}. \end{aligned}\right\} \tag{6.4.36}$$

Electron synchrotrons normally operate at energies which are large compared with the rest-mass energy; the initial acceleration is effected by the betatron mechanism. Hence, during the synchrotron mode of operation, $\mathsf{p} \approx \mathsf{e}$ and $\mathsf{e} \gg 1$; we now see from (6.4.13) that $\lambda \approx a$, so that the frequency of the accelerating field is almost constant, and that (6.4.31), (6.4.34) and (6.4.36) reduce to

$$\left.\begin{aligned} |y| \propto \mathsf{e}^{-\frac{1}{2}}, \quad |x_B| \propto \mathsf{e}^{-\frac{1}{2}}, \quad |x_S| \propto \mathsf{e}^{-\frac{3}{4}},\\ |\phi_S^{(1)}| \propto \mathsf{e}^{-\frac{1}{4}}, \quad |e_S^{(1)}| \propto \mathsf{e}^{\frac{1}{4}}. \end{aligned}\right\} \tag{6.4.37}$$

6.5. Perturbation calculations

In this section we shall establish a method of calculating the influence upon trajectories of terms which appear in the variational function but are not taken into account in the paraxial treatment. We shall assume that these non-linear terms are small compared with those which appear in the paraxial or linear treatment, and we shall confine ourselves to first-order calculations. Since the non-linear terms involve higher powers of the off-axis co-ordinates than the linear terms, they will be comparatively small provided that the off-axis co-ordinates of trajectories remain bounded within small limits. However, the primary object of perturbation calculations is indeed to discover whether effects which are not taken into account by the linear theory will have the effect of increasing these co-ordinates beyond any limit—a process sometimes referred to as 'blow-out'. It is clear, therefore, that one can in any real instance give only a qualified answer to this problem; one can only hope to say 'my estimates of the effects of these or those terms are *consistent* with the *hypothesis* of stability' (i.e. of bounded trajectories), for it is always possible that still higher terms of the variational function will lead to progressive expansion of the beam.‡ This is a fundamental shortcoming of theories of particle accelerators which has no counterpart in theories of static electron-optical instruments; one can *ensure* that the beam of an electron microscope will not exceed a certain diameter by the appropriate choice and disposition of no more than two diaphragms.

It was stated in the introduction to this chapter that one is also interested in the effect of constructional errors upon trajectories. Such errors will entail a deviation of the variational function from its ideal, part of which, as in the problem of §4.6, will appear in the first- and second-order terms of the variational function. These effects also one would wish to examine by perturbation methods.

It is proposed that, to maintain consistency with earlier chapters, we attempt only a first-order treatment of perturbations. It must be pointed out, however, that although first-order treatments are usually adequate in the discussion of static electron-optical problems, first-order calculations are frequently inadequate for the study of particle accelerators. If, for instance, the next terms in the expansion of the variational function beyond the paraxial term were of the *fourth* order, a first-order perturbation calculation would yield an estimate of the change in period of oscillations; but if the next terms were of the *third* order, one would need to make a second-order perturbation calculation in order to obtain an estimate of the change in period of oscillations.

It would be possible to obtain the formulae we seek by starting from the formulae we have established for perturbation characteristic functions in §3.5, but this approach will be deferred for use in the next chapter. In the present section we shall proceed along lines which are clearly parallel to those of §3.5, but we shall assume from

‡ There appears to be a close analogy here with the difficulty of 'small divisors' of planetary theory. See, for instance, E. T. Whittaker, *A History of the Theories of Aether and Electricity* (Nelson, London, 1953), vol. 2, p. 146.

the start that our unperturbed variational function is independent of the co-ordinate z.

We see from (6.2.15) and (6.2.20) that if all the arguments except z of the variational function are subjected to arbitrary variation, the resulting variation of the function itself satisfies the differential relation

$$\delta M = p_i' \delta x_i + p_\tau' \delta \tau + p_i \delta x_i' + p_\tau \delta \tau'. \tag{6.5.1}$$

Now suppose that the *function* $M(x_i, \tau, x_i', \tau', z)$ is perturbed thus:

$$M \to M + \Delta M; \tag{6.5.2}$$

we shall work to the first order only in quantities of the order of magnitude of ΔM. This perturbation of M causes a perturbation of the trajectories, so that, if for the time being we confine our attention to a length of the accelerator which is short enough for the perturbed trajectories to diverge only slightly from the unperturbed trajectories, we may write

$$x_i \to x_i + \Delta x_i, \quad p_i \to p_i + \Delta p_i, \quad \text{etc.} \tag{6.5.3}$$

Since the arguments of the variational function will be affected by (6.5.3), we must distinguish between the *functional* perturbation, which we have written as (6.5.2), and the total perturbation which we shall denote by $\Delta_t M$. Since we are working to the first order only, $\Delta_t M$ is the sum of two parts, one due to the change in the function and the other due to the change in its arguments. The latter may be evaluated by means of (6.5.1), so that

$$\Delta_t M = \Delta M + p_i' \Delta x_i + p_\tau' \Delta \tau + p_i \Delta x_i' + p_\tau \Delta \tau'. \tag{6.5.4}$$

If we now write down the first-order contribution to the perturbed form of the differential relation (6.5.1),

$$\delta \Delta_t M = \Delta p_i' \delta x_i + p_i' \delta \Delta x_i + \Delta p_\tau' \delta \tau + p_\tau' \delta \Delta \tau + \Delta p_i \delta x_i' + p_i \delta \Delta x_i' + \Delta p_\tau \delta \tau' + p_\tau \delta \Delta \tau', \tag{6.5.5}$$

and then make use of (6.5.4), we obtain the *first-order perturbation relation*

$$\delta \Delta M = (\Delta p_i' \delta x_i + \Delta p_\tau' \delta \tau + \Delta p_i \delta x_i' + \Delta p_\tau \delta \tau') - (\Delta x_i' \delta p_i + \Delta \tau' \delta p_\tau + \Delta x_i \delta p_i' + \Delta \tau \delta p_\tau'). \tag{6.5.6}$$

Although differential relations of this type were convenient for application to problems in static electron optics, they are not suitable for application to problems concerning particle accelerators since even if the perturbation of the variational function is small the trajectories contemplated are so long that the perturbed trajectory cannot be supposed to remain close to the unperturbed trajectory indefinitely. We therefore proceed as follows:

Suppose that the trajectory equations for the unperturbed system have been integrated completely so that the off-axis co-ordinates and their conjugate momenta are known as functions of z and of six parameters which we write as α_m:

$$x_i = x_i(\alpha_m; z), \quad p_i = p_i(\alpha_m; z), \quad \text{etc.} \tag{6.5.7}$$

We may take account of the perturbation by retaining the functional form (6.5.7) for the trajectories but allowing α_m to vary (slowly if ΔM is small) with z. If we compare a perturbed and an unperturbed trajectory which, at some arbitrary value of z, have the same co-ordinates and momenta and hence the same values of α_m, we see that

$$\Delta x_i = 0, \quad \Delta p_i = 0, \quad \text{etc.}, \tag{6.5.8}$$

but

$$\Delta x_i' = \alpha_m' \frac{\partial x_i}{\partial \alpha_m}, \quad \Delta p_i' = \alpha_m' \frac{\partial p_i}{\partial \alpha_m}, \quad \text{etc.}, \tag{6.5.9}$$

where α_m' is the rate of change of α_m in the perturbed system. Hence the differential relation (6.5.6) reduces to

$$\delta \Delta M = \{\alpha_m, \alpha_n\} \alpha_m' \delta \alpha_n, \tag{6.5.10}$$

from which it follows that if ΔM is expressed as a function of the α_m by replacing its arguments x_i and τ by the functions (6.5.7), the variation of these parameters due to the perturbation is determined by the differential equations

$$\frac{d\alpha_m}{dz} \{\alpha_m, \alpha_n\} = \frac{\partial \Delta M}{\partial \alpha_n}, \tag{6.5.11}$$

where the Lagrange brackets are now defined explicitly as

$$\{\alpha, \beta\} = \frac{\partial p_i}{\partial \alpha} \frac{\partial x_i}{\partial \beta} + \frac{\partial p_\tau}{\partial \alpha} \frac{\partial \tau}{\partial \beta} - \frac{\partial p_i}{\partial \beta} \frac{\partial x_i}{\partial \alpha} - \frac{\partial p_\tau}{\partial \beta} \frac{\partial \tau}{\partial \alpha}. \tag{6.5.12}$$

Although the formula (6.5.11) is appropriate for the evaluation of the effect of non-linear terms of the variational function in a uniform-focusing accelerator, it would not be appropriate for the evaluation of the effect of periodic disturbances such as arise in a circular accelerator due to constructional inaccuracies. In this case it is best to integrate the equation (6.5.11) over one revolution so as to obtain

$$\Delta\alpha_m\{\alpha_m, \alpha_n\} = \frac{\partial \Delta U}{\partial \alpha_n}, \tag{6.5.13}$$

where now

$$\Delta U = \int \Delta M \, dz, \tag{6.5.14}$$

the integral being taken over the appropriate period of the independent co-ordinate z. We are assuming that the parameters α_m change only slightly over one period so that they may be given constant values inside the integral. We shall see in § 7.4 that (6.5.13) is exactly the formula one obtains if one starts from the perturbation characteristic function and its differential relation, but the later treatment will be more general since we shall not assume from the outset that M is independent of z.

It is worth noticing that since the Lagrange brackets which appear in (6.5.11) and (6.5.13) are invariants of the unperturbed system, they may be expected to vary only slowly under the perturbation.

As an example of the application of the foregoing theory, let us consider the following:

$$\left.\begin{aligned} M &= \tfrac{1}{2}\mathsf{p}x'^2 - \tfrac{1}{2}\mathsf{p}\omega_x^2 x^2, \\ \Delta M &= \varpi \mathsf{p}\omega_x^2 \cos\omega_z z \,.\, x^2, \end{aligned}\right\} \tag{6.5.15}$$

where ϖ is supposed to be small and positive. We shall find in the next section that such terms represent the effect of the phase oscillations upon the radial oscillations in a linear accelerator. The *exact* trajectory equation derivable from $M + \Delta M$ will be recognized as the Mathieu equation,‡ so that we shall be able to check the veracity of our results.

The solution of the unperturbed system is

$$\left.\begin{aligned} x &= K\sin(\omega_x z + \kappa), \\ p_x &= \mathsf{p}\omega_x K\cos(\omega_x z + \kappa) \end{aligned}\right\} \tag{6.5.16}$$

so that

$$\{K, \kappa\} = \mathsf{p}\omega_x K; \tag{6.5.17}$$

moreover, ΔM may be expressed as

$$\Delta M = \varpi \mathsf{p}\omega_x^2 K^2 \cos\omega_z z \sin^2(\omega_x z + \kappa). \tag{6.5.18}$$

Application of (6.5.11) now yields the equations

$$\kappa' = \tfrac{1}{2}\varpi\omega_x[2\cos\omega_z z - \cos\{(2\omega_x + \omega_z) z + 2\kappa\} - \cos\{(2\omega_x - \omega_z) z + 2\kappa\}] \tag{6.5.19}$$

‡ See, for instance, N. W. MacLachlan, *Theory and Application of Mathieu Functions* (Oxford University Press, 1947).

and

$$K^{-1}K' = -\tfrac{1}{2}\varpi\omega_x[\sin\{(2\omega_x+\omega_z)z+2\kappa\} + \sin\{(2\omega_x-\omega_z)z+2\kappa\}]. \quad (6.5.20)$$

It is clear from these equations that if none of ω_z and $2\omega_x \pm \omega_z$ is small, the term 2κ which appears in the arguments of the circular functions cannot appreciably affect the behaviour of $\kappa(z)$ and $K(z)$, so that the latter must oscillate with small amplitudes near their initial values; hence there is in this case no instability.

However, let us now consider the interesting case for which

$$\omega_x = \omega + \Delta\omega, \quad \omega_z = 2\omega, \quad (6.5.21)$$

where $\Delta\omega$ is supposed to be small. The dominant parts of the equations (6.5.19) and (6.5.20) are then

$$\kappa' = -\tfrac{1}{2}\varpi\omega\cos(2\Delta\omega\, z + 2\kappa) \quad (6.5.22)$$

and

$$K^{-1}K' = -\tfrac{1}{2}\varpi\omega\sin(2\Delta\omega\, z + 2\kappa). \quad (6.5.23)$$

If we make the substitution

$$\psi = 2\Delta\omega\, z + 2\kappa \quad (6.5.24)$$

so that (6.5.22) and (6.5.23) become

$$\psi' = 2\Delta\omega - \varpi\omega\cos\psi \quad (6.5.25)$$

and

$$K^{-1}K' = -\tfrac{1}{2}\varpi\omega\sin\psi, \quad (6.5.26)$$

we can see how the solution depends upon $\Delta\omega$ and ϖ.

If $|\Delta\omega| > \tfrac{1}{2}\varpi\omega$, $|\psi'| \geqslant 2|\Delta\omega| - \varpi\omega > 0$, so that the phase angle 'rotates' continually. The differential equation

$$\frac{1}{K}\frac{dK}{d\psi} = -\frac{1}{2}\frac{\varpi\omega\sin\psi}{2\Delta\omega - \varpi\omega\cos\psi}, \quad (6.5.27)$$

which may be formed from (6.5.25) and (6.5.26), may then be integrated to give

$$K = C|\Delta\omega - \tfrac{1}{2}\varpi\omega\cos\psi|^{-\frac{1}{2}}, \quad (6.5.28)$$

from which we see that the amplitude K is bounded but oscillates. The ratio of the maximum and minimum values taken by K is given by

$$\frac{K_{\text{max.}}}{K_{\text{min.}}} = \sqrt{\left(\frac{|\Delta\omega| + \tfrac{1}{2}\varpi\omega}{|\Delta\omega| - \tfrac{1}{2}\varpi\omega}\right)}, \quad (6.5.29)$$

from which we see that this ratio tends to infinity as $\Delta\omega$ approaches the critical value $\tfrac{1}{2}\varpi\omega$. The increase is fairly slow however; for instance, if $\Delta\omega$ has twice its critical value, the ratio of maximum to minimum is only $\sqrt{3}$ (see fig. 6.5). The distance occupied by a complete cycle is found from (6.5.25) to be given by

$$\pi/\sqrt{[(\Delta\omega)^2 - \tfrac{1}{4}\varpi^2\omega^2]}, \tag{6.5.30}$$

so that the oscillations become slower and slower as the critical value of $\Delta\omega$ is approached.

If, on the other hand, $|\Delta\omega| < \frac{1}{2}\varpi\omega$, it is easy to see that ψ tends to a limiting position. We see from (6.5.25) that there are two

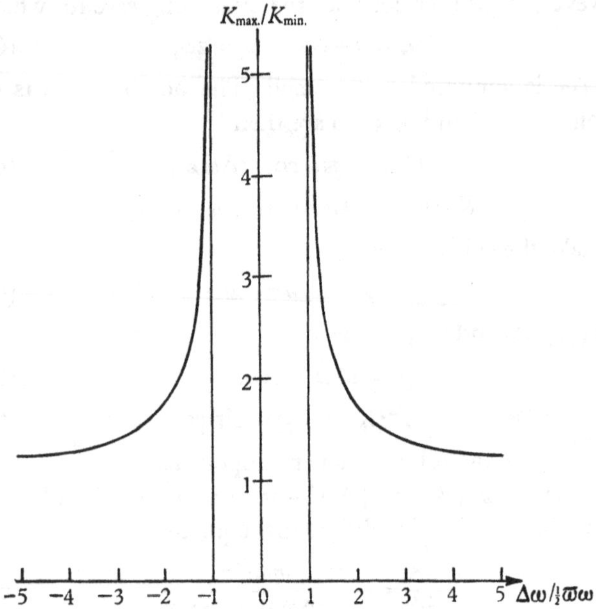

Fig. 6.5. Variation of amplitude near a band of instability.

stationary positions of ψ: one is a limiting position for z increasing but the other is a limiting position for z decreasing. The former is defined by

$$\cos\psi_0 = \Delta\omega/\tfrac{1}{2}\varpi\omega, \quad \pi < \psi_0 < 2\pi, \tag{6.5.31}$$

and we find from (6.5.25) that a small departure from this value decays exponentially with 'relaxation length' given by

$$L_R = 1/\sqrt{[\varpi^2\omega^2 - 4(\Delta\omega)^2]}. \tag{6.5.32}$$

Hence a small displacement from the stable position of ψ decreases by a factor of e in a distance L_R. The limiting behaviour of K as this stable position is approached is readily seen to be

$$K = K_0 \exp\{z/\sqrt{[\tfrac{1}{4}\varpi^2\omega^2 - (\Delta\omega)^2]}\}, \qquad (6.5.33)$$

which represents a monotonic expansion of the beam.

We may sum up by saying that, according to first-order perturbation theory, the perturbation produces a band of instability centred about $\frac{1}{2}\omega_z$ and of half-width $\frac{1}{2}\varpi\omega_z$. As this band is approached the amplitude oscillates; quickly at first, then more slowly but more violently. Within the band there is an exponential expansion.

6.6. The linear electron accelerator

As a second example of a 'uniform-focusing' accelerator we select the linear accelerator in the form in which it is used for the acceleration of electrons. We saw in § 6.4 that it is possible, in the synchrotron, to achieve focusing in both off-axis directions and in phase simultaneously, since the former is provided by the deflecting magnetic field and the latter is ensured by the existence of a stable phase angle. There is no deflecting field in the linear accelerator, and it is found that the accelerating field itself cannot provide both phase focusing and radial focusing; it is therefore necessary to provide a static focusing field, and this usually takes the form of an axial magnetic field. In accordance with the fundamental approximation of this chapter, we assume that the accelerating field and the static magnetic field vary only slowly along the length of the accelerator, but it must be emphasized that, although our treatment will bring out the principal dynamical properties of the linear accelerator, fields which can be realized in practice differ significantly from those of our ideal.

It is proposed that, following the pattern of Chapter 4, we adopt rectangular co-ordinates (x, y, z) with the z-axis along the axis of symmetry of the accelerator, but compound x and y into one complex co-ordinate,

$$u = x + iy. \qquad (6.6.1)$$

The static magnetic field may then be represented by the complex potential which was introduced in § 4.2; from (4.2.10),

$$U_{\text{mag.}} = i(\tfrac{1}{2}u\Psi - \tfrac{1}{16}\bar{u}u^2\Psi'' + \ldots), \qquad (6.6.2)$$

where $\Psi(z)$ is the magnetic scalar potential upon the axis.

If we follow the suggestion of § 6.4 and represent the accelerating (time-varying) field by a vector potential distribution, we see that

(6.4.6) again enables us to introduce a complex potential. It is not difficult to see from (4.2.7) that (6.4.7) will be satisfied if we require that U should satisfy the equation

$$\nabla^2 U - \partial^2 U/\partial t^2 = 0. \tag{6.6.3}$$

The series expansion (4.2.10) must therefore be replaced by the expansion

$$U = \tfrac{1}{2}U_1 u - \tfrac{1}{16}\left(\frac{\partial^2 U_1}{\partial z^2} - \frac{\partial^2 U_1}{\partial t^2}\right)\bar{u}u^2 + \dots, \tag{6.6.4}$$

where U_1 is expressible as $U_1(z, t)$. Moreover, it follows from (4.2.7) and (6.4.8) that the field upon the z-axis is related to U_1 by the equations

$$E_z = -\frac{\partial U_{1r}}{\partial t}, \quad H_z = \frac{\partial U_{1i}}{\partial z}, \tag{6.6.5}$$

where it has been expedient to separate U_1 into the real and imaginary parts,

$$U_1 = U_{1r} + \mathrm{i}U_{1i}. \tag{6.6.6}$$

Now the accelerating field in a linear accelerator is periodic in time and is designed to produce a purely electric field along the axis of symmetry. We may therefore write

$$E_z(z, t) = -\mathsf{E}\cos\left(\int_0^z k(\zeta)\,d\zeta - \lambda^{-1}t\right) \tag{6.6.7}$$

for the field on the axis, where $2\pi\lambda$ is the free-space wavelength of the accelerating electromagnetic field and $1/\lambda k(z)$ is the phase velocity as a function of z. We shall of course expect to make the phase velocity equal to the electron velocity. It follows from (6.6.5) and (6.6.7) that the accelerating field should be represented by a function $U_1(z, t)$ which is purely real and may be taken to be

$$U_1 = -\mathsf{E}\sin\left(\int_0^z k(\zeta)\,d\zeta - \lambda^{-1}t\right). \tag{6.6.8}$$

The complex potential representing both the magnetic and the accelerating fields is obtained by adding (6.6.2) to (6.6.4) and adopting (6.6.8).

Since the curvature of the axis vanishes, since there is no electric scalar potential, and since we are adopting complex co-ordinates and representing the field by a complex potential, we should write (6.2.16) in the form

$$M = -\sqrt{(T'^2 - 1 + 2T'\tau' + \tau'^2 + \bar{u}'u')} + \frac{1}{2}\left(\bar{u}'\frac{\partial U}{\partial z} + u'\frac{\partial \bar{U}}{\partial z}\right) - \left(\frac{\partial \bar{U}}{\partial \bar{u}} + \frac{\partial U}{\partial u}\right). \tag{6.6.9}$$

In order to evaluate (6.6.9) explicitly, we first express U as a function of z, t, $\bar{u}$ and u and then carry out the transformation (6.2.12). We next assume that the electron describing the trajectory axis rides at a fixed position on the wave of the accelerating field, i.e. we write

$$\int_0^z k(\zeta)\,d\zeta - \lambda\!\!\!^-{}^{-1}T(z) \equiv \phi. \tag{6.6.10}$$

It follows at once from (6.2.25) that $k(z)$ is related to the particle energy by

$$k = \mathsf{p}^{-1}(1+\mathsf{e})\lambda\!\!\!^-{}^{-1}. \tag{6.6.11}$$

The linear part of M is found to be

$$M^{(1)} = -(1+\mathsf{e})\tau' - \mathsf{E}\cos\phi\,.\,\tau. \tag{6.6.12}$$

If we now define a distance L by

$$d\mathsf{e}/dz = L^{-1}, \tag{6.6.13}$$

so that L *is the distance in which an electron will increase its energy by one rest mass*, it follows from (6.6.12) that

$$L^{-1} = \mathsf{E}\cos\phi. \tag{6.6.14}$$

If we now evaluate the quadratic part of M and use the above relations, we obtain

$$M^{(2)} = \tfrac{1}{2}\mathsf{p}^3\tau'^2 - \tfrac{1}{2}\lambda\!\!\!^-{}^{-1}L^{-1}\tan\phi\,.\,\tau^2 + \tfrac{1}{2}\mathsf{p}\bar{u}'u' + \tfrac{1}{4}i\mathsf{H}(\bar{u}'u - \bar{u}u') + \tfrac{1}{4}\mathsf{p}^{-2}\lambda\!\!\!^-{}^{-1}L^{-1}\tan\phi\,.\,\bar{u}u, \tag{6.6.15}$$

where H is the strength of the magnetic field upon the axis. The familiar transformation (4.2.15),

$$u = v\exp\left(\tfrac{1}{2}i\int_0^z \mathsf{p}^{-1}(\zeta)\,\mathsf{H}(\zeta)\,d\zeta\right), \tag{6.6.16}$$

now yields

$$M^{(2)} = \tfrac{1}{2}\mathsf{p}^3\tau'^2 - \tfrac{1}{2}\lambda\!\!\!^-{}^{-1}L^{-1}\tan\phi\,.\,\tau^2 + \tfrac{1}{2}\mathsf{p}\bar{v}'v' - \tfrac{1}{8}(\mathsf{p}^{-1}\mathsf{H}^2 - 2\mathsf{p}^{-2}\lambda\!\!\!^-{}^{-1}L^{-1}\tan\phi)\,\bar{v}v, \tag{6.6.17}$$

from which we may derive the paraxial approximation to the trajectory equations:

$$\frac{d^2\tau}{dz^2} + \mathsf{p}^{-3}\lambda\!\!\!^-{}^{-1}L^{-1}\tan\phi\,.\,\tau = 0, \tag{6.6.18}$$

$$\frac{d^2v}{dz^2} + \tfrac{1}{4}(\mathsf{p}^{-2}\mathsf{H}^2 - 2\mathsf{p}^{-3}\lambda\!\!\!^-{}^{-1}L^{-1}\tan\phi)\,v = 0. \tag{6.6.19}$$

It is clear from (6.6.13) and (6.6.14) that in order to have acceleration we must have $\cos\phi > 0$, and we see from (6.6.18) that in order to have phase stability $\tan\phi$ must be positive. Hence the phase-stable position for particles is just in front of the crest of the accelerating electric wave; this is as we should expect, for a particle behind this position will receive an increased driving force and a particle ahead of this position a diminished force. We also see from (6.6.19) that in order to have radial focusing, the magnetic field strength must exceed some critical value; we must have

$$\mathsf{H}^2 > 2\mathsf{p}^{-1}\lambda^{-1}L^{-1}\tan\phi \equiv \mathsf{H}_c^2. \qquad (6.6.20)$$

It is interesting to note from this relation that the magnetic field strength may be allowed to *decrease* along the accelerator since, provided the electric field strength and stable phase angle and hence also L remain constant, the critical value of H varies with the electron energy as $\mathsf{p}^{-\frac{1}{2}}$. The solutions of the paraxial trajectory equations are simply

$$\tau = K_\tau \sin(\omega_\tau z + \kappa_\tau) \qquad (6.6.21)$$

and $$r = K_r \sin(\omega_r z + \kappa_r), \qquad (6.6.22)$$

where $$\omega_\tau^2 = \mathsf{p}^{-3}\lambda^{-1}L^{-1}\tan\phi \qquad (6.6.23)$$

and $$\omega_r^2 = \tfrac{1}{4}(\mathsf{p}^{-2}\mathsf{H}^2 - 2\mathsf{p}^{-3}\lambda^{-1}L^{-1}\tan\phi), \qquad (6.6.24)$$

and where r may be taken to denote either component of v or the radial displacement of an electron which happens to cross the axis.

The variation of the amplitudes of the phase and radial oscillations may be obtained from the formulae given on p. 165. Since the momentum conjugate to r is $\mathsf{p}r'$,

$$|r| \propto \mathsf{p}^{-\frac{1}{2}}\omega_r^{-\frac{1}{2}}; \qquad (6.6.25)$$

since the momentum conjugate to τ is $\mathsf{p}^3\tau'$,

$$|\phi^{(1)}| \propto \mathsf{p}^{-\frac{3}{4}}, \qquad (6.6.26)$$

where we have noted from (6.6.10) that the phase increment is related to τ by $$\phi^{(1)} = -\lambda^{-1}\tau. \qquad (6.6.27)$$

We see from (6.6.24) and (6.6.25) that if the magnetic field strength is chosen to be some constant times the critical value of the field strength, $|r| \propto \mathsf{p}^{\frac{1}{4}}$, so that the beam expands slowly along the accelerator. If, on the other hand, the field were strong and maintained a constant value along the accelerator, the beam would maintain a constant diameter (as is, of course, obvious when we remember that electrons would then move along the magnetic lines of force). We note from the last paragraph of § 6.3 that, for maximum acceptance, ω_r should be made large at injection and $\mathsf{p}\omega_r$ should be kept constant or made to increase with z. The amplitude of the phase oscillations decreases along the accelerator.

On evaluating the third-order contribution to the variational function, we obtain

$$M^{(3)} = -\tfrac{1}{2}\mathsf{p}^4(1+\mathsf{e})\tau'^3 + \tfrac{1}{6}\lambda\!\!\!^-{}^{-2}L^{-2}\tau^3 - \tfrac{1}{2}\mathsf{p}^2(1+\mathsf{e})\tau'\bar{v}'v' + \tfrac{1}{4}\mathsf{p}^{-1}(1+\mathsf{e})\lambda\!\!\!^-{}^{-1}L^{-1}\tan\phi\,\tau'\bar{v}v - \tfrac{1}{4}\mathsf{p}^{-2}\lambda\!\!\!^-{}^{-2}L^{-1}\tau\bar{v}v. \quad (6.6.28)$$

The terms of this function which involve τ and τ' only will affect only the phase oscillations; a first-order perturbation analysis of the effect of these terms would show a shift in the average phase of a trajectory from the stable phase position by an amount proportional to the square of the amplitude of the phase oscillation. A second-order perturbation analysis of these terms would yield a change in the frequency of the phase oscillations by an amount proportional to the square of the amplitude (but so also would a first-order perturbation treatment of the relevant terms of $M^{(4)}$).

A complete first-order perturbation treatment of the remaining terms of $M^{(3)}$ would show that there is a band of instability centred at $\omega_r = \tfrac{1}{2}\omega_\tau$, and that in this region a large phase amplitude would give rise to expansion of the radial amplitude and vice versa. Since normally the permissible radial amplitude in a linear accelerator is severely limited by the wave-guide structure,‡ it is proposed that we now make the simplifying assumption that the phase oscillations are unaffected by the coupling terms in (6.6.28) and proceed to investigate the effect of these terms on the radial oscillations. This effect has already been studied in the last section. (It is not difficult to see that the results we established are not significantly altered if we replace $\omega_x^2 x^2$ in the expression for ΔM in (6.5.15) by x'^2.) We saw that the principal effect of terms of the former type is to introduce a band of instability centred about $\omega_r = \tfrac{1}{2}\omega_\tau$. If the magnetic field is to be strong enough to effect radial focusing but weak enough to satisfy $\omega_r < \tfrac{1}{2}\omega_\tau$, it must lie within the following limits:

$$\mathsf{H}_c < \mathsf{H} < \sqrt{\tfrac{3}{2}}\,\mathsf{H}_c. \quad (6.6.29)$$

An alternative is, of course, to specify that $\mathsf{H} > \sqrt{\tfrac{3}{2}}\,\mathsf{H}_c$.

‡ See, for instance, ref. (48).

In order to determine the width of the band of instability, it is necessary to look more closely into the terms of (6.6.28). Of the last three terms, the ratio of the magnitudes of the first to the second is found to be $2(\omega_r/\omega_\tau)^2$ so that, if (6.6.29) is satisfied, the first is less important than the second. The ratio of the magnitudes of the second to the third term is $\mathsf{p}^{-\frac{1}{2}}(1+\mathsf{e})(\lambda\!\!\!^{-}/L)^{\frac{1}{2}}\tan^{\frac{3}{2}}\phi$; since the wavelength of the accelerating field is normally small compared with the distance in which an electron acquires an energy of one rest mass, $\lambda\!\!\!^{-}/L$ is small, so that the third term will be more important than the second except at very low and at very high energies. We may therefore make an estimate of the band width by taking into account only the last term of (6.6.28).

We find by comparing (6.5.15) with (6.6.17) and (6.6.28) and noting that, under the conditions of interest, $\omega_r \approx \frac{1}{2}\omega_\tau$, the appropriate value of ϖ is given by

$$\varpi = |\phi^{(1)}| \cot\phi. \tag{6.6.30}$$

The width of the band of instability is, we remember, $\frac{1}{2}\varpi\omega_\tau$, from which we find that H must differ from $\sqrt{\frac{3}{2}}\,\mathsf{H}_c$ by more than $\frac{1}{4}\varpi\mathsf{H}_c$. However, the instability which this band represents is not disastrous; the maximum possible logarithmic rate of expansion may be found from (6.5.33) to be $\frac{1}{2}\pi|\phi^{(1)}|\cot\phi$ per cycle of phase oscillation, so that if $\phi = \frac{1}{4}\pi$ and $\phi^{(1)} = 0{\cdot}2$ (corresponding to an accepted width of 20°), the beam diameter will expand by a factor of three in three complete cycles of phase oscillation. Since an electron executes very few oscillations in a linear accelerator, we may infer that the beam would not suffer a serious expansion even if there were instability of the type we are considering over a short length of the accelerator.

It is interesting to consider the numerical magnitude of the quantities for which we have derived formulae by giving a practical example. If we consider an accelerator designed to accelerate electrons with initial energy of 50 keV to a final energy of 8 MeV in 4 metres, travelling at a stable phase angle of 45° and employing an electric field with a free-space wavelength of 10 cm., we see that $\lambda\!\!\!^{-} = 1{\cdot}6$ cm., $L = 50$ cm., $\mathsf{E} = 0{\cdot}55$ and $\phi = \frac{1}{4}\pi$, and that initially $\mathsf{e} = 0{\cdot}1$ and $\mathsf{p} = 0{\cdot}45$, finally $\mathsf{e} = 8$ and $\mathsf{p} = 9$. We see at once from (6.6.26) that the band of phase angle covered by a bunch of electrons will diminish by a factor of 10 during acceleration, so that if the accelerator initially accepts a band of 50°, the final width will be only 5°. Initially and finally, ω_τ has the values 0·37 and 0·0041 corresponding to wavelengths of 17 cm. and 15½ metres, respectively, for the phase oscillations.‡ The final energy spread in the beam may be found from (6.2.27):

$$|e^{(1)}| = \mathrm{p}^3\omega_\tau\lambda\!\!\!^{-}{}^{-1}|\phi^{(1)}|; \qquad (6.6.31)$$

if, as above, $|\phi^{(1)}|$ is finally 0·05, the half-width of the energy spread is found to be 120 keV. The critical value of the magnetic field strength is found from (6.6.20) to fall off from 400 to 90 gauss; if the magnetic field strength is chosen to be some constant multiple of its critical value, we see from (6.6.24) and (6.6.25) that the beam will expand to twice its initial radius during acceleration. Under these conditions, the wavelength of the radial oscillations will be comparable with the wavelength of the phase oscillations.

‡ It is worth mentioning that since oscillations are very slow at the end of a linear accelerator, the use of the adiabatic approximation is open to some criticism, since the kinetic energy of the beam will vary considerably during the period of a phase or radial oscillation.

CHAPTER 7

PERIODIC FOCUSING IN PARTICLE ACCELERATORS

7.1. Introduction

It was observed at the end of the introduction to the preceding chapter that although one can carry out a simplified analysis of the dynamical properties of certain particle accelerators by approximating the electromagnetic fields to 'uniform' fields which vary slowly and smoothly along the length of the accelerator, one would wish to be able to make more realistic studies by taking into account the periodic distribution of accelerating and focusing fields. Such a procedure, which is not necessary for discussion of the linear electron accelerator in which the periodicity of the fields (i.e. of the wave-guide structure) is small compared with the periodicity of phase and radial oscillations but might be thought desirable for machines such as the synchrotron in which the periodicity of the accelerating fields (i.e. of the resonators), is comparable with that of the betatron oscillations, is indispensable for the theoretical study of the 'race track',‡ in which the focusing field is far removed from uniformity, or of the 'microtron',§ in which electrons gain one rest-mass unit of energy per revolution. The appropriate procedure was seen by LeCouteur|| in his investigation of the 'regenerative deflector' (an extraction mechanism used in the synchrocyclotron) to be the matrix method adopted by Floquet¶ in his study of linear differential equations with periodic coefficients.

However, the decisive impetus to the development of the analysis which is to be set out in this chapter came in 1953 when Courant, Livingston and Snyder‡‡ discovered that the adoption of periodic rather than uniform focusing in the proton synchrotron was so advantageous as to increase by a factor of about ten the energy to which particles could be accelerated for given capital outlay. We saw in §6.4 that in the uniform-focusing synchrotron about one

‡ Refs. (50) and (51). § Refs. (52) and (53). || Ref. (54).

¶ See, for instance, E. T. Whittaker and G. N. Watson, *A Course of Modern Analysis* (Cambridge University Press, 4th ed. 1952), p. 412.

‡‡ Ref. (55).

betatron oscillation occurs per revolution; in a proton synchrotron which has a radius measured in hundreds of feet, the diameter of cross-section of the vacuum chamber must be measured in feet if the angular divergence of the beam is to be more than a fraction of a degree so that the deflecting magnet must be correspondingly massive. It was found that if the magnet were divided into a large

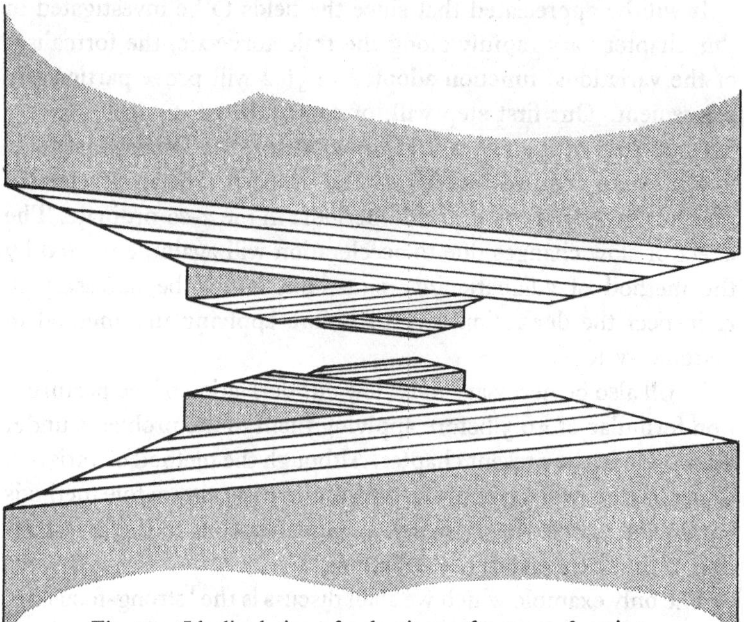

Fig. 7.1. Idealized view of pole-pieces of a strong-focusing proton synchrotron looking along the trajectory axis.

number of sectors, each with a large field gradient $\partial H/\partial r$, the gradient alternating in sign as indicated in fig. 7.1, stability in both off-axis directions could be attained; indeed, the frequency of betatron oscillations could be increased considerably so that the same angular beam acceptance could be obtained with a vacuum chamber of a few inches cross-section, thus reducing significantly the dimensions of the deflecting magnet which is the most expensive component of such a machine.

We see from (6.4.19) and the first of equations (6.4.23), or equivalently from equations (5.4.10), that if the field gradient is large and is made to alternate in sign, the focusing in either of the off-axis directions can be represented by a sequence of converging and diverging thick lenses. The achievement of Courant and his collaborators was to discover (or, rather, to rediscover for themselves‡) that such a sequence can have a strong net converging action and then to appreciate the significance of this fact. This method of focusing will be referred to as 'strong focusing', the term first suggested by its proponents.

It will be appreciated that since the fields to be investigated in this chapter vary rapidly along the trajectory axis, the formalism of the variational function adopted in § 6.2 will prove particularly convenient. Our first step will, of course, be to suppose that the paraxial part of the variational function is of the orthogonal form (6.2.29), and then to investigate the properties of an accelerator which is supposed to be strictly periodic in the z-co-ordinate. The effect of slow changes due to acceleration will again be treated by the method of adiabatic invariants, but it will be necessary to re-inspect the derivation of § 6.2 before applying this method to periodic systems.

It will also be necessary to review our derivation of the perturbation formulae of § 6.5 before applying them to the problems under discussion in the present chapter. Although the method of variation of parameters will again prove useful, there are alternative methods employing matrix theory which are advantageous in the investigation of linear (paraxial) perturbations.

The only example which we shall discuss is the 'strong-focusing' synchrotron proposed by the Brookhaven team and, for simplicity, we shall confine our attention to the betatron oscillations which may be treated by the methods of static electron optics since (as we saw in § 6.4) the accelerating field plays no part in the mechanism of these oscillations. The most interesting aspect of this theory will prove to be the calculation of the effect upon the stability of the machine of various types of constructional error, such as incorrect machining and alinement of the pole-pieces and of non-linear terms in the trajectory equations.

‡ The focusing mechanism which is now known as 'strong-focusing' was in fact patented by Nicholas Christofilos in 1950 and 1951 (U.S. Patents, nos. 2,531,028 and 2,567,904). (Courant, E. D., Livingston, M. S., Snyder, H. S. and Blewett, J. R., *Phys. Rev.* 91 (1953), 202.) Mr Christofilos attempted to interest the Berkeley team in his idea but met with no success, since his unorthodox mathematics was difficult to follow (see *Scientific American*, **188** (1953), no. 6, 45–6).

7.2. The paraxial theory

The basic assumption of this chapter is that the variational function M introduced in §6.2 is periodic in the independent variable z (although we shall also take into account, in §7.3, an aperiodic slow variation due to acceleration). In the present section we shall further assume that the paraxial part of the variational function may be expressed in the orthogonal form (6.2.29), so that in our first analysis of the paraxial properties of periodic-focusing accelerators we need consider only one off-axis co-ordinate. It is proposed that we set up along the length of the trajectory axis a sequence of 'reference planes', separated by a distance equal to the wavelength of the periodicity, which we shall number 1, 2, 3, etc.; the accelerator is in this way broken down into a sequence of identical dynamical systems.

If we consider the off-axis co-ordinate x and its conjugate momentum p_x, we see that the values of these variables in any one plane are determined uniquely by their values in any other plane. In particular, since we are working to the paraxial approximation, we may write the relation between the values taken by these variables in adjacent planes in matrix form as

$$\begin{pmatrix} p_{x,k+1} \\ x_{k+1} \end{pmatrix} = \begin{pmatrix} A & B \\ C & D \end{pmatrix} \begin{pmatrix} p_{x,k} \\ x_k \end{pmatrix}, \tag{7.2.1}$$

where the coefficients A, B, C, D must, by our hypothesis, be independent of k. This relation is a restatement of (3.4.22) which referred to static electron-optical systems.

It is convenient to have an alternative notation for matrix expressions such as (7.2.1). We may write (7.2.1) as‡

$$v_{i,k+1} = M_{ij} v_{j,k} \tag{7.2.2}$$

if

$$\begin{pmatrix} v_1 \\ v_2 \end{pmatrix} = \begin{pmatrix} p_x \\ x \end{pmatrix}, \quad \begin{pmatrix} M_{11} & M_{12} \\ M_{21} & M_{22} \end{pmatrix} = \begin{pmatrix} A & B \\ C & D \end{pmatrix}. \tag{7.2.3}$$

If we take the one-dimensional form of the Lagrange bracket (6.5.12) and then consider $\{p_{x,k}, x_k\}$, we obtain once more the relation (3.4.23), i.e.

$$AD - BC = 1, \tag{7.2.4}$$

‡ No confusion should arise between M the matrix and M the variational function.

so that the matrix M_{ij} has unit determinant

$$\| M_{ij} \| = 1. \tag{7.2.5}$$

Since the equation (7.2.1) is linear in p_x and x, the general solution of this iterative relation may be expressed as a linear combination of any two different particular solutions. If both these particular solutions have bounded variations, then the general trajectory has bounded variation and the system is 'convergent', to use an optical term, or 'stable' to use a dynamical one; if, on the other hand, one of the particular solutions has unbounded variation, only an infinitesimal number of the particles making up a beam can be expected to remain within a limited distance of the trajectory axis, so that the system must be considered 'divergent' or 'unstable'.

The particular solutions which we shall seek are those which are expressible as

$$v_{i,k} = \lambda^k v_i^{(e)}. \tag{7.2.6}$$

The vector $v_i^{(e)}$ is an 'eigen-vector' or 'characteristic vector' corresponding to the 'eigen-value' or 'characteristic root' λ of the matrix M_{ij}. We see from (7.2.2) that

$$(M_{ij} - \lambda\delta_{ij})\, v_j^{(e)} = 0, \tag{7.2.7}$$

but this equation determines non-zero vectors only if the determinant of the appropriate matrix is zero, i.e. if

$$\| M_{ij} - \lambda\delta_{ij} \| = 0. \tag{7.2.8}$$

The condition (7.2.8), i.e.

$$\begin{vmatrix} A-\lambda & B \\ C & D-\lambda \end{vmatrix} = 0, \tag{7.2.9}$$

may be written explicitly as

$$\lambda^2 - (A+D)\,\lambda + 1 = 0 \tag{7.2.10}$$

by virtue of (7.2.4). We see from (7.2.10) that if λ is a root, so are $\bar{\lambda}$, the complex conjugate, $1/\lambda$ and $1/\bar{\lambda}$. The roots must therefore be expressible either as one of the pairs $e^{\pm\mu}$ and $-e^{\pm\mu}$, or as the pair $e^{\pm i\theta}$, where μ and θ are real. It is easy to see that if

$$|A+D| > 2, \tag{7.2.11}$$

then $$\lambda = e^{\mu},\ e^{-\mu} \quad \text{or} \quad -e^{\mu},\ -e^{-\mu}, \tag{7.2.12}$$

where $$2\cosh\mu = A+D \quad \text{or} \quad -(A+D), \tag{7.2.13}$$

respectively. If, on the other hand,

$$|A+D|<2, \tag{7.2.14}$$

then

$$\lambda = e^{i\theta},\ e^{-i\theta}, \tag{7.2.15}$$

where

$$2\cos\theta = A+D. \tag{7.2.16}$$

On referring to (7.2.6), we now see that if (7.2.11) is satisfied, the system is certainly divergent and that if (7.2.14) is satisfied, the system is certainly convergent. The angle θ defined by (7.2.15) might be termed the 'characteristic phase' of the system.

The case $|A+D|=2$, which is excluded from (7.2.11) and (7.2.14), calls for special attention. If $A+D$, which is known as the 'spur' or 'trace' of the matrix appearing in (7.2.1), has the value 2, and if A and D are not equal to unity, it follows from (7.2.4) that M_{ij} is expressible as

$$(M_{ij}) = \begin{pmatrix} 1+\alpha & -\alpha\beta \\ \alpha/\beta & 1-\alpha \end{pmatrix}, \tag{7.2.17}$$

where α and β are necessarily finite and non-zero. The general solution of the relation (7.2.1) is now of the form

$$\begin{pmatrix} p_{x,k} \\ x_k \end{pmatrix} = a\begin{pmatrix} \beta \\ 1 \end{pmatrix} + b\begin{pmatrix} -k\alpha\beta \\ 1-k\alpha \end{pmatrix}, \tag{7.2.18}$$

so that the system is divergent. The same observation holds if $A+D=-2$ and $A \neq D$. If $A=D=\pm 1$, we see from (7.2.4) that either B or C must vanish. It is not difficult to see, or to prove from the appropriate limiting forms of (7.2.18), that if $B=0$ but $C \neq 0$, x_k increases linearly with k although $p_{x,k}$ is constant; and that if $C=0$ but $B \neq 0$, $p_{x,k}$ increases linearly although x_k is constant. Both these cases are divergent. This leaves only the possibility $A=D=\pm 1$, $B=C=0$, which clearly is convergent.

We may sum up by saying that *the system represented by the iterative relation* (7.2.1) *is convergent if and only if the spur of the matrix has absolute value less than two or the matrix is either the unit matrix or its negative.* In the remainder of this section we shall suppose that (7.2.14) is satisfied.

Since the 'eigen-solutions' of the relation (7.2.1) are expressible as (7.2.6), where λ is given by (7.2.15), the general real solution must be expressible as

$$\left.\begin{aligned} p_{x,k} &= \rho K \sin(k\theta+\kappa+\psi), \\ x_k &= K\sin(k\theta+\kappa), \end{aligned}\right\} \tag{7.2.19}$$

where K and κ are parameters. By inserting (7.2.19) into (7.2.1) and noting from (7.2.4) and (7.2.14) that B and C must have opposite signs, we find that the coefficients ρ and ψ may be taken to be given by

$$\rho = \sqrt{(-B/C)} \tag{7.2.20}$$

and

$$\cos\psi = \tfrac{1}{2}(A-D)/\sqrt{(-BC)} \quad (0<\psi<\pi); \tag{7.2.21}$$

we may also verify that

$$\sin\psi = |\sin\theta|/\sqrt{(-BC)}. \tag{7.2.22}$$

Now we see from (3.4.28) that C measures, in a certain sense, the 'optical length' of a section; let us therefore specify that $\sin\theta$ should have the same sign as C and that $-\pi<\theta<\pi$. The elements of the matrix may now be expressed in terms of θ, ψ and ρ; we find that

$$\left.\begin{aligned} A &= \frac{\sin(\theta+\psi)}{\sin\psi}, & B &= -\frac{\rho\sin\theta}{\sin\psi}, \\ C &= \frac{\sin\theta}{\rho\sin\psi}, & D &= -\frac{\sin(\theta-\psi)}{\sin\psi}. \end{aligned}\right\} \tag{7.2.23}$$

There is a certain combination of ρ and ψ which will appear frequently, and it is therefore proposed that we write

$$\sigma = \rho\sin\psi; \tag{7.2.24}$$

we see from (7.2.23) that σ may also be expressed as

$$\sigma = C^{-1}\sin\theta. \tag{7.2.25}$$

It was implied at the beginning of this section that the sections of the accelerator which are characterized by the matrix M_{ij} or, equivalently, by the coefficients θ, ψ and ρ, should be the irreducible sections of which the accelerator is composed. In a *circular* accelerator, however, great importance attaches to the matrix characterizing a complete revolution, which will comprise a number of sections, since, whatever errors arise in the construction of the accelerator, the focusing field represented by this matrix is 'reproduced' exactly. If there are N sections to a revolution, we may deduce from (7.2.19) that the matrix characterizing a revolution is determined by the coefficients Θ, ψ and ρ, where ψ and ρ are the same as for a single section, but

$$\Theta = N\theta. \tag{7.2.26}$$

It is proposed that we ignore integral multiples of 2π and so consider that Θ lies in the same range as θ, i.e. that $-\pi<\Theta<\pi$.

We may also obtain the inverse of the matrix M_{ij} by replacing θ in (7.2.23) by $-\theta$; thus

$$(M_{ij})^{-1} = \begin{pmatrix} D & -B \\ -C & A \end{pmatrix}. \tag{7.2.27}$$

Let us now consider the general form of matrices which represent *symmetrical* sections, that is, sections whose focusing fields are unaltered on reflexion in the plane which lies midway between the two relevant reference planes. On remembering that if, for a given assembly of trajectories, z is changed to $-z$, the sign of p_x is changed, we see that the matrix representing a reflected section may be obtained from (7.2.27) by changing the signs attached to B and C. Hence *for a symmetrical section* $A = D$, *and so* $\psi = \frac{1}{2}\pi$. We now see from (7.2.23) that the matrix representing a symmetrical section is expressible simply as

$$(M_{ij})_{\text{symm.}} = \begin{pmatrix} \cos\theta & -\rho\sin\theta \\ \rho^{-1}\sin\theta & \cos\theta \end{pmatrix}. \tag{7.2.28}$$

The conclusions of the above paragraph suggest that one may on occasion wish to change from one sequence of reference planes to another; if these sets are to have the same periodicity, they can differ only by a constant displacement. Let us suppose that the co-ordinates of the two sequences are z_k and $\tilde{z}_k$; then $\tilde{z}_k - z_k$ is independent of k. If the matrices representing the sections of accelerator between z_k and z_{k+1} and between $\tilde{z}_k$ and $\tilde{z}_{k+1}$ are M_{ij} and $\tilde{M}_{ij}$, respectively, we see that

$$\tilde{M}_{ik} S_{kj} = S_{ik} M_{kj}, \tag{7.2.29}$$

where S_{ij} is the matrix representing the section of accelerator between the planes at z_k and $\tilde{z}_k$ (or at z_{k+1} and $\tilde{z}_{k+1}$). Hence

$$\tilde{M}_{ij} = S_{ik} M_{kl} S^{-1}_{lj}. \tag{7.2.30}$$

We may draw an important conclusion from the form of (7.2.30): since $\tilde{M}_{ii} = M_{ii}$, the spur of the matrix has not been changed by the change in the sequence of reference planes. Hence *the characteristic exponents are independent of one's choice of reference planes.* One may also see this result by noting that the spur of the product of two matrices is unchanged if the order of these matrices is reversed. It must be emphasized that both ψ and ρ will in general be changed by a displacement of reference planes.

Let us now consider the 'strong-focusing' system in its simplest form. We consider a system of period $2L$, so that $z_{k+1}-z_k=2L$, whose variational function is given by

$$\left.\begin{aligned} M &= \tfrac{1}{2}\mathrm{p}x'^2-\tfrac{1}{2}\mathrm{p}\omega^2x^2 \quad z_k<z<z_k+L, \\ &= \tfrac{1}{2}\mathrm{p}x'^2+\tfrac{1}{2}\mathrm{p}\omega^2x^2 \quad z_k+L<z<z_{k+1}. \end{aligned}\right\} \tag{7.2.31}$$

We easily solve the trajectory equations (6.2.15) and so obtain the relations

$$\begin{pmatrix} p_x(z_k+L) \\ x(z_k+L) \end{pmatrix} = \begin{pmatrix} \cos\omega L & -p\omega\sin\omega L \\ p^{-1}\omega^{-1}\sin\omega L & \cos\omega L \end{pmatrix} \begin{pmatrix} p_x(z_k) \\ x(z_k) \end{pmatrix} \tag{7.2.32}$$

and

$$\begin{pmatrix} p_x(z_{k+1}) \\ x(z_{k+1}) \end{pmatrix} = \begin{pmatrix} \cosh\omega L & p\omega\sinh L \\ p^{-1}\omega^{-1}\sinh\omega L & \cosh\omega L \end{pmatrix} \begin{pmatrix} p_x(z_k+L) \\ x(z_k+L) \end{pmatrix}. \tag{7.2.33}$$

On multiplying the matrices appearing in the above two equations, we find that the elements of the matrix in (7.2.1) are given by

$$\left.\begin{aligned} A &= \cos\omega L\cosh\omega L+\sin\omega L\sinh\omega L, \\ B &= \mathrm{p}\omega(\cos\omega L\sinh\omega L-\sin\omega L\cosh\omega L), \\ C &= \mathrm{p}^{-1}\omega^{-1}(\sin\omega L\cosh\omega L+\cos\omega L\sinh\omega L), \\ D &= \cos\omega L\cosh\omega L-\sin\omega L\sinh\omega L. \end{aligned}\right\} \tag{7.2.34}$$

Hence

$$A+D=2\cos\omega L\cosh\omega L. \tag{7.2.35}$$

The variation of $A+D$ with ωL is shown in fig. 7.2. It is seen that for values of ωL between 0 and 1·88 (108°) the system is convergent; thereafter the system is divergent except for narrow bands centred at the values $L=\tfrac{1}{2}(2n+1)\pi$. The choice $\omega L=\tfrac{1}{2}\pi$, which is sometimes referred to as the '$\tfrac{1}{2}\pi$ mode' of operation, is particularly simple for the matrix coefficients (7.2.34) then reduce to

$$\left.\begin{aligned} A &= \sinh\tfrac{1}{2}\pi, & B &= -\mathrm{p}\omega\cosh\tfrac{1}{2}\pi, \\ C &= \mathrm{p}^{-1}\omega^{-1}\cosh\tfrac{1}{2}\pi, & D &= -\sinh\tfrac{1}{2}\pi \end{aligned}\right\} \tag{7.2.36}$$

so that, from (7.2.16), (7.2.20), (7.2.21), (7.2.22) and (7.2.24),

$$\left.\begin{aligned} \theta=\tfrac{1}{2}\pi, \quad \rho=\mathrm{p}\omega, \quad \psi=\arccos(\tanh\tfrac{1}{2}\pi) \quad (\approx 23\tfrac{1}{2}^\circ), \\ \sigma\approx 0{\cdot}4\mathrm{p}\omega. \end{aligned}\right\} \tag{7.2.37}$$

We have already seen that if a section is 'reflected', A and D are interchanged; hence if the order of the convergent and divergent fields which constitute a section is reversed, θ and ρ are unchanged but ψ is replaced by $\pi-\psi$, that is, in the above example, by $156\frac{1}{2}°$; σ is unchanged.

There are two ways in which we may displace the reference planes in the system defined by (7.2.31) in order to make sections symmetrical. If the planes are moved forward by a distance $\frac{1}{2}L$, so that each section contains a diverging field sandwiched between two converging fields, we find that, for the $\frac{1}{2}\pi$ mode,

$$(M_{ij})_{CDC}=\begin{pmatrix} \cdot & -\mathrm{p}\omega\, e^{-\frac{1}{2}\pi} \\ \mathrm{p}^{-1}\omega^{-1}e^{\frac{1}{2}\pi} & \cdot \end{pmatrix} \tag{7.2.38}$$

so that $$\theta=\tfrac{1}{2}\pi, \quad \psi=\tfrac{1}{2}\pi, \quad \sigma=\rho=e^{-\frac{1}{2}\pi}\mathrm{p}\omega\approx 0{\cdot}21\,\mathrm{p}\omega. \tag{7.2.39}$$

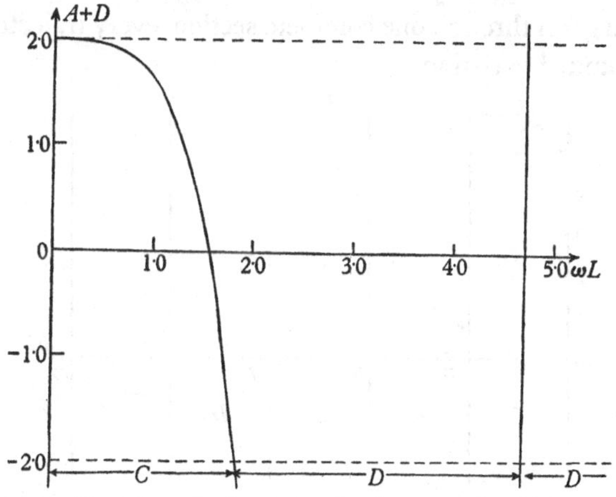

Fig. 7.2. Focusing criterion for a simple strong-focusing system.

If, on the other hand, the planes are moved backwards by a distance $\frac{1}{2}L$ so that a converging field is sandwiched between two diverging fields, we find that

$$(M_{ij})_{DCD}=\begin{pmatrix} \cdot & -\mathrm{p}\omega \\ \mathrm{p}^{-1}\omega^{-1} & \cdot \end{pmatrix}, \tag{7.2.40}$$

so that $$\theta=\tfrac{1}{2}\pi, \quad \psi=\tfrac{1}{2}\pi, \quad \sigma=\rho=\mathrm{p}\omega. \tag{7.2.41}$$

It was implied in the introduction to this chapter, and we shall find explicitly in §7.8, that in the strong-focusing bevatron a field which is convergent in the x-direction is divergent in the y-direction and vice versa. The above two paragraphs now show that by appropriate choice of the reference planes one may reduce the difference between the x and y matrices to either a difference in phase angle ψ or a difference in the coefficient ρ.

Although the preceding mathematical considerations demonstrate that a sequence of converging and diverging fields of equal strengths will, if the appropriate condition is satisfied, exhibit a strong net focusing effect, the reader will doubtless wish to form a picture of this rather surprising process. Fig. 7.3 shows a trajectory in a strong-focusing system operating in the $\frac{1}{2}\pi$ mode, and it is seen that, although the amplitude of the trajectory increases considerably in the divergent section, the convergent section is nevertheless strong enough to reverse the slope of the trajectory. Since a linearly independent trajectory can be formed by displacing that shown in the diagram through one complete section, every trajectory must have bounded variation.

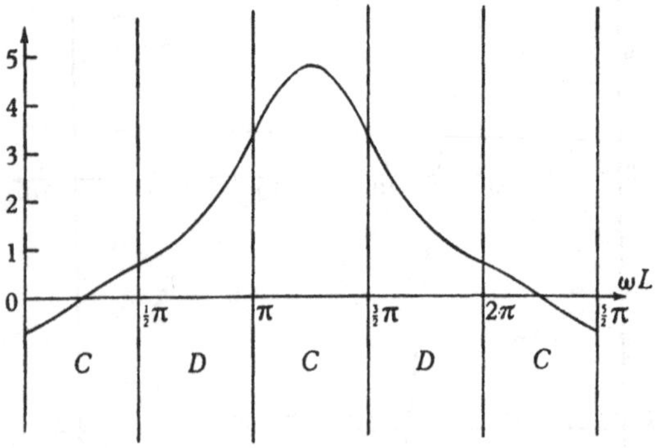

Fig. 7.3. Trajectory in a strong-focusing system operating in the $\frac{1}{2}\pi$ mode.

7.3. The adiabatic invariants

We now return to the consideration of adiabatic invariants which were introduced in §6.3 in order to determine under what conditions these may be employed in the theory of particle accelerators with periodic focusing. In order to retrace the argument of the above-mentioned section, we again adopt time as independent variable and again introduce the Hamiltonian for the system. In place of (6.3.1), however, we must write

$$\mathscr{H} = \mathscr{H}(x_r, p_s, t; c), \tag{7.3.1}$$

where we suppose the above function to be periodic in the variable t with period $2\pi\omega^{-1}$. Similarly, the functions on the right-hand side of (6.3.3) must be assumed to be periodic in their third variable with period $2\pi\omega^{-1}$.‡

No modification is called for in the argument of 'phase independence' leading up to (6.3.7), but modifications will be necessary in the justification of this hypothesis. Since the equations (6.3.2) are still valid, so are the equations (6.3.8). But now note that while $\{\alpha_m, \alpha_n\}^{-1}$ is still slowly varying, $\{\alpha_n, c\}$ is periodic in both $\omega_i(c)\,t+\kappa_i$ and in $\omega(c)\,t$ (with period 2π in each case). It is therefore necessary to carry out a multiple Fourier expansion in the arguments $\omega_i(c)\,t+\kappa_i$ together with $\omega(c)\,t$, from which we see immediately that in order for the integral (6.3.10) to be independent of the phases κ_i we must retain the conditions (*a*), (*b*) and (*c*) of p. 162, and add to these the further condition: (*d*) *ω should not be expressible as a linear combination of the ω_i with rational coefficients.*

There is no difficulty in expressing the above result in terms of the formalism introduced in §6.2 and developed in §7.2 for application to periodic-focusing accelerators. If, as is appropriate, the contour of integration is chosen to lie in a 'reference plane', we obtain once more the formula (6.3.14), i.e.

$$\mathsf{A}_l = \oint d\kappa_l \left(p_i \frac{\partial x_i}{\partial \kappa_l} + p_r \frac{\partial \tau}{\partial \kappa_l}\right) \quad (l \text{ n.t.b.s.}), \qquad (7.3.2)$$

in which the further suffix k, denoting the particular reference plane adopted, is understood. However, the co-ordinates and momenta entering in (7.3.2) will now take values, along a general trajectory, given by formulae similar to (7.2.19) rather than (6.3.16) and (6.3.17). Hence the arguments $\omega_i t+\kappa_i$ will now be replaced by arguments $k\theta_i+\kappa_i$, and the periodicity which was represented by the argument ωt is now represented by our sequence of reference planes. Hence we see that, in the present formalism, condition (*a*) and the further condition (*d*) necessary for the existence of adiabatic invariants in periodic-focusing accelerators may be combined into

‡ The existence of periodic solutions of equations of motion with periodic coefficients is admittedly problematical except in the linear approximation. Nevertheless, we lose nothing and incur no complication by proceeding with the general case, and it should be noted that although no analytical demonstration has been given, computational investigations suggest that such periodic solutions sometimes do exist (see, for instance, ref. (52)). We have here one of the most challenging mathematical problems related to particle accelerators.

the statement that *no linear combination of the* θ_i *with rational coefficients should be identically equal to zero or* 2π *for all* k. Once again the condition is much less stringent in the paraxial approximation, for it may be shown, by an argument similar to that in §6.3, that in this case it is sufficient to specify that no θ_i is an integral multiple of π and no $\theta_i \pm \theta_j$ $(i \neq j)$ is an integral multiple of 2π.

The reader may care to look at the above result in another way. The relation embodied in (6.3.10) may be expressed in terms of 'sections' and 'reference planes' by dividing the integral into an appropriate number of parts. Thus we obtain

$$\alpha_{mb} - \alpha_{ma} = \Sigma\{\alpha_m, \alpha_n\}^{-1}\,\delta c\,\overline{\{\alpha_n, c\}}, \tag{7.3.3}$$

where the summation is over the appropriate range of values of k, which is implicitly present in (7.3.3), δc is the change in c between reference planes, and

$$\overline{\{\alpha_n, c\}} = \int \mathrm{d}t\{\alpha_n, c\} \Big/ \int \mathrm{d}t, \tag{7.3.4}$$

the integrations here being over one section. The Lagrange bracket $\{\alpha_m, \alpha_n\}$ is again slowly varying and c is still assumed aperiodic, but the term $\overline{\{\alpha_n, c\}}$ is periodic, with characteristic exponents θ_i, and may therefore be expanded as a multiple Fourier series in $k\theta_i + \kappa_i$. It is now clear that if any linear combination of the θ_i with integral coefficients were an integral multiple of 2π, one of the terms of the expansion would cause the sum (7.3.4) to contain a term which grows steadily, the value of this term depending upon the phases κ_i. The simplification to the above condition appropriate to the paraxial approximation may be verified by noting that in this case $\overline{\{\alpha_n, c\}}$ is a quadratic combination of circular functions with arguments $k\theta_i + \kappa_i$.

It was stated in §6.3 that although adiabatic invariants could be expressed in the familiar form of time integrals taken over one period of oscillation (cf. (6.3.13)) when the system is singly periodic and time-independent, such a formulation is not possible for a Hamiltonian varying periodically with time; consequently adiabatic invariants which arise in the study of periodic-focusing accelerators cannot be expressed in this form. We may now see why this is so. In reducing the general formula (6.3.12) to the particular form

(6.3.13) we used the facts (i) that γ_3, whose equation is $p_r = \text{const.}$, $x_r = \text{const.}$, lies in the tube of trajectories under consideration; and (ii) that the total energy $-p_\epsilon$ is constant over the entire tube (see fig. 6.4). In the present section, however, we are assuming that the Hamiltonian depends explicitly upon time, from which we see that neither of the above conditions is now satisfied.

The simplest but most useful example of the application of the adiabatic invariant to periodic-focusing accelerators may be obtained by assuming that the paraxial system considered in § 7.2, which involves only one off-axis co-ordinate, is slowly varying. If the integral (7.3.2) is evaluated for the trajectory (7.2.19) we obtain, with the notation of (7.2.24), the expression

$$\mathsf{A} = \pi\sigma K^2 \tag{7.3.5}$$

as the relevant adiabatic invariant. We see from (7.2.25) and (7.3.5) that the variation of the amplitude of x in the reference planes during acceleration is given by

$$|x| \propto \sqrt{\left|\frac{C}{\sin\theta}\right|}. \tag{7.3.6}$$

If, for example, the fields are so designed that the x-focusing remains in the $\frac{1}{2}\pi$ mode throughout the acceleration, we see from (7.2.34) and (7.3.6) that the amplitude of the trajectories varies as $\mathsf{p}^{-\frac{1}{2}}\omega^{-\frac{1}{2}}$. It is also interesting to note that (7.3.6) shows how the amplitude of a given trajectory in the reference planes varies with one's choice of reference planes. Since θ is unaffected by a displacement of the sequence, the amplitude is proportional to $|C|^{\frac{1}{2}}$. We may, for instance, deduce from (7.2.38) and (7.2.40) that, in the $\frac{1}{2}\pi$ mode, the amplitude of an assembly of trajectories with constant K varies by a factor of $e^{\frac{1}{4}\pi}$ in passing from the centre of a divergent field to the centre of a convergent field (cf. fig. 7.3).

Since the expression (7.3.5) is quadratic in K, we may expect it to be expressible as a quadratic combination of p_x and x. The coefficients of this quadratic form may be found by substituting the expressions (7.2.19) for p_x and x and then giving κ three different values in turn. In this way we obtain the alternative formula

$$\mathsf{A} = \pi \operatorname{cosec}\theta (Cp_x^2 - (A-D)p_x x - Bx^2). \tag{7.3.7}$$

That the expression enclosed in brackets on the right-hand side of (7.3.7) is an invariant (though not an adiabatic invariant) may be seen in the following way: Since we are working to the paraxial approximation, we may deduce from the Lagrange differential invariant (2.3.2) that if $p_{x,k}^{(1)}$, $x_k^{(1)}$ and $p_{x,k}^{(2)}$, $x_k^{(2)}$ are the co-ordinates of any two trajectories,

$$p_{x,k}^{(1)}x_k^{(2)} - p_{x,k}^{(2)}x_k^{(1)} = \text{const.} \quad (k \text{ n.t.b.s.}). \tag{7.3.8}$$

If we now suppose that a second trajectory is formed from an arbitrary trajectory simply by displacing it through the length of one section, we find that the invariant (7.3.8) is identical with the expression enclosed in brackets in (7.3.7). Since trajectories which may be formed by further displacements are linear combinations of the first two, further invariants formed in this way must be multiples of the invariant (7.3.7).

The invariant (7.3.7) offers further insight into the nature of the condition (7.2.14) for convergence. If we ignore for the time being the possibility of adiabatic perturbation, the quadratic expression in (7.3.7) is constant for any trajectory, so that the point $(p_{x,k}, x_k)$ in the 'phase space' p_x-x must move (with k) upon the contour, which will be a conic section, whose equation is of the form (7.3.7). Hence trajectories will be bounded, and the system convergent, if the quadratic form is positive-definite or negative-definite so that the contour in the p_x-x plane is an ellipse; the system is otherwise divergent (except that the case $A=D=1$, $B=C=0$ calls for special investigation and has been found to be convergent). The condition that the quadratic form in (7.3.7) should be of definite sign is that

$$(A-D)^2+4BC<0 \tag{7.3.9}$$

which, by virtue of (7.2.4), is equivalent to (7.2.14).

Let us now consider the acceptance of an accelerator with periodic focusing. If the trajectories are bounded by apertures of width $2a$ situated at the reference planes, we see from (7.3.5) that the area of the p_x-x plane which is accepted by the accelerator is given by

$$\pi a^2 \min(\sigma_k) \tag{7.3.10}$$

which corresponds, let us say, to a value K_0 of K upon injection. It is interesting now to find the shape of the 'acceptance envelope', and this may be obtained as the curve along which the expression on the right-hand side of (7.3.7) takes the value (7.3.10). We find with the help of (7.2.23) that the equation of this envelope may be written in the form

$$\frac{(x_0+\mathrm{p}_0\rho_0^{-1}x_0')^2}{4K_0^2\cos^2\frac{1}{2}\psi_0}+\frac{(x_0-\mathrm{p}_0\rho_0^{-1}x_0')^2}{4K_0^2\sin^2\frac{1}{2}\psi_0}=0, \qquad (7.3.11)$$

which is clearly an ellipse.

The significance of (7.3.11) may be given more practical form by determining the pair of virtual apertures near the injection plane, supposed for this purpose to be in field-free space, whose acceptance envelope most nearly fits (7.3.11). An aperture of width $2a$ situated

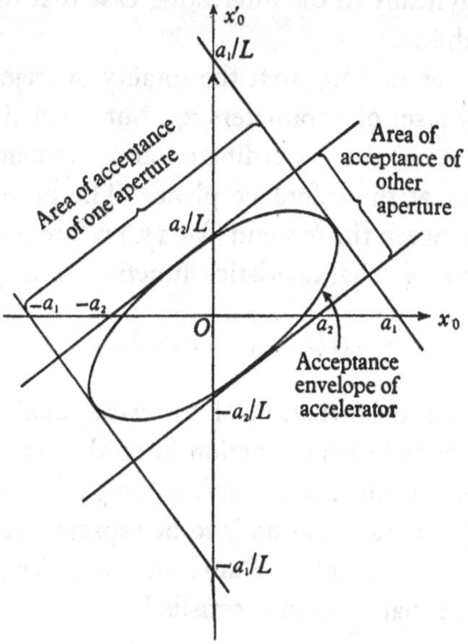

Fig. 7.4. Approximation of acceptance characteristics by a pair of apertures.

a distance L on the accelerator side of the injection plane imposes the inequality

$$|x_0+Lx_0'|<a \qquad (7.3.12)$$

upon the values of x_0 and x_0'. Hence the pair of apertures whose acceptance envelope is the rectangle of closest fit of the ellipse (7.3.11) (see fig. 7.4) have positions and half-widths given by

$$\left.\begin{aligned} L_1 &= \mathrm{p}_0\rho_0^{-1}, & a_1 &= 2K_0\cos\tfrac{1}{2}\psi_0, \\ L_2 &= -\mathrm{p}_0\rho_0^{-1}, & a_2 &= 2K_0\sin\tfrac{1}{2}\psi_0. \end{aligned}\right\} \qquad (7.3.13)$$

If the sections are symmetrical, so that $\psi_0=\frac{1}{2}\pi$, these apertures are of course symmetrically disposed about the entrance plane.

7.4. Perturbation calculations by variation of parameters

We shall establish in this section generalizations of the formulae of § 6.5 which will enable us to solve problems involving perturbations of periodic-focusing accelerators by the method of variation of parameters. These formulae, which may be obtained with ease by drawing on the perturbation theory of static electron optics, offer a convenient method for computing the effects of perturbations. Moreover, as we saw in § 6.5, such equations may sometimes be integrated analytically in the interesting case that the perturbation produces instability.

We suppose, as in § 6.5, that the totality of trajectories may be enumerated by a set of parameters α_m, but we shall need to know the dependence of the co-ordinates and momenta upon these parameters only at the reference planes. The perturbation of the section lying between the kth and $(k+1)$th reference planes may be represented by a characteristic function $\Delta U_k(p_{ik}, p_{\tau k}, x_{ik}, \tau_k)$ defined by

$$\Delta U_k = \int_{z}^{z_{k+1}} \Delta M \, dz, \tag{7.4.1}$$

where ΔM is the perturbation of the variational function. This perturbation characteristic function is of the type introduced in § 2.4, and we see from (2.4.17) and a comparison of (2.4.19) and (2.4.20) that if the perturbation is to be represented by an *effective* perturbation of p_{ik}, x_{ik}, etc., of amounts Δp_{ik}, Δx_{ik}, etc., then the following differential relation is satisfied:

$$\delta \Delta U_k = (\Delta p_{ik} \delta x_{ik} + \Delta p_{\tau k} \partial \tau_k) - (\Delta x_{ik} \delta p_{ik} + \Delta \tau_k \delta p_{\tau k}) \quad (k \text{ n.t.b.s.}). \tag{7.4.2}$$

We note from this relation that

$$\Delta p_{ik} = \frac{\partial \Delta U_k}{\partial x_{ik}}, \quad \Delta p_{\tau k} = \frac{\partial \Delta U_k}{\partial \tau_k}, \quad \Delta x_{ik} = -\frac{\partial \Delta U_k}{\partial p_{ik}}, \quad \Delta \tau_k = -\frac{\partial \Delta U_k}{\partial p_{\tau k}} \quad (k \text{ n.t.b.s.}). \tag{7.4.3}$$

The relation (7.4.2) may be re-expressed in terms of the increments of the parameters α_m as

$$\delta \Delta U_k = \Delta \alpha_{mk} \{\alpha_m, \alpha_n\} \delta \alpha_{nk} \quad (k \text{ n.t.b.s.}), \tag{7.4.4}$$

where $\{\alpha_m, \alpha_n\}$ is the Lagrange bracket defined by (6.5.12). It follows immediately from (7.4.4) that the change in the values of the parameters over any section due to the perturbation in that section may be obtained from the set of simultaneous equations

$$\Delta\alpha_{mk}\{\alpha_m, \alpha_n\}_k = \frac{\partial \Delta U_k}{\partial \alpha_{nk}} \quad (k \text{ n.t.b.s.}). \qquad (7.4.5)$$

It is here assumed that the integral (7.4.1) taken, as usual, along an unperturbed trajectory, is expressed as a function of the parameters α_m. The relation (7.4.5) is identical with (6.5.13), but it has now been established for sections of a periodic-focusing accelerator whereas in §6·5 it was shown to hold only for sections of a uniform-focusing accelerator.

As an example of the application of the above method, it is interesting to consider briefly the periodic-focusing analogue of the problem treated in §6.5, i.e. the effect of coupling between the phase and radial motions in a strong-focusing linear accelerator. It will suffice to consider only one off-axis co-ordinate, x; then, to the paraxial approximation, the variations of this co-ordinate and of the phase co-ordinate τ are of the form (7.2.19). Since, as we saw in §6.6, the lowest-order coupling term of the variational function is linear in the phase co-ordinate and quadratic in the radial co-ordinate, the perturbation characteristic function for a typical section must be expressible as a linear combination of $p_{\tau k}$ and τ_k times a quadratic combination of p_{xk} and x_k. On referring to (7.2.19) we see that such a function must have the form

$$\Delta U = K_\tau K_x^2\{L\sin(k\theta_\tau + \kappa_\tau + \lambda) + M\sin(k(2\theta_x + \theta_\tau) + 2\kappa_x + \kappa_\tau + \mu) + N\sin(k(2\theta_x - \theta_\tau) + 2\kappa_x - \kappa_\tau + \nu)\}, \qquad (7.4.6)$$

where the suffix k has been suppressed. Since, from (7.2.19) and (7.2.24),

$$\{K, \kappa\} = \sigma K, \qquad (7.4.7)$$

the complete set of difference equations (7.4.5) derivable from (7.4.6) is

$$\left.\begin{aligned} \Delta K_x &= 2\sigma_x^{-1}K_\tau K_x(M\cos(\ ) + N\cos(\ )),\\ \Delta\kappa_x &= -2\sigma_x^{-1}K_\tau^{-1}K_x^2(L\sin(\ ) + M\sin(\ ) + N\sin(\ )),\\ \Delta K_\tau &= \sigma_\tau^{-1}K_x^2(L\cos(\ ) + M\cos(\ ) - N\cos(\ )),\\ \Delta\kappa_\tau &= -\sigma_\tau^{-1}K_\tau^{-1}K_x^2(L\sin(\ ) + M\sin(\ ) + N\sin(\ )), \end{aligned}\right\} \qquad (7.4.8)$$

in which the arguments of the circular functions are understood to be those of (7.4.6). Such a set of equations, it will be noted, is well suited for numerical solution.

Since in a linear accelerator one is concerned to prevent radial expansion of the beam, one may begin by assuming that K_x in (7.4.8) is small; hence we may neglect the variations of K_τ and κ_τ, regarding these as constants. This is the approximation implicit in § 6.5. We now see from the first of equations (7.4.8) that radial expansion may occur if $2\theta_x \pm \theta_\tau$ is nearly an integral multiple of 2π. Without going into details, we may see that the method adopted in § 6.5 may be repeated. We may write $\theta_x = \frac{1}{2}\theta_\tau + \epsilon$, where ϵ is to be small, and write the argument of the circular function associated with N as ψ. The first two equations of (7.4.8) may now be expressed as difference equations for ψ and $\log K_x$ which may be solved in the same way as the differential equations (6.5.25) and (6.5.26). We find that if ϵ lies outside a certain band, the system is stable, but that as ϵ approaches the edge of the band the amplitude of oscillations begins to fluctuate more and more violently; within the band there is exponential expansion.

Inspection of (7.4.8) shows that another interesting case arises when θ_τ is close to zero; if θ_x is not near to $\pm\pi$, there will in this instance be no radial expansion, so that we may relax our earlier restriction that both K_τ and κ_τ are effectively constant in order to investigate whether there will be an expansion of the phase oscillations. Only the last two equations of (7.4.8) are then significant, for K_x will remain constant since the first of equations (7.4.8) does not involve the term in L.

If we write $\theta_\tau = \epsilon$ and assume that ϵ is small and also write

$$\psi = k\epsilon + \kappa_\tau + \lambda \tag{7.4.9}$$

and

$$\varpi = \sigma_\tau^{-1} K_x^2 L, \tag{7.4.10}$$

the important contributions to the last two equations of (7.4.8) reduce to

$$\left.\begin{aligned} \Delta K_\tau &= \varpi \cos\psi, \\ \Delta\psi &= \epsilon - \varpi K_\tau^{-1} \sin\psi. \end{aligned}\right\} \tag{7.4.11}$$

We see that the steady solution of (7.4.11) is

$$K_\tau^S = \varpi/\epsilon, \quad \psi = \tfrac{1}{2}\pi \tag{7.4.12}$$

and it may be proved that in the general solution K_τ oscillates about K_τ^S. This shows that for small enough ϵ, i.e. for θ_τ close enough to zero, the phase amplitude will grow to large values. It is not suggested that the particular instability which we have considered would be important in a proton linear accelerator, since the period of the accelerating field (i.e. of the drift-tubes) is much smaller than that of the phase oscillations, so that the perturbation coefficient ϖ is necessarily very small. Nevertheless, the example is instructive in demonstrating that *in approximating a periodic-focusing accelerator to a uniform-focusing model, one runs the risk of neglecting certain of the potential instabilities of the real machine.*

7.5. Perturbation calculations by matrix methods

The method of variation of parameters, which was considered in the previous section, is useful in the consideration of non-linear perturbations (thus the complete set of equations (7.4.8) would yield the solution to a non-linear problem) or in the discussion of linear problems in which the periodicity represented by the sequence of reference planes is violated (the particular case that $2\theta_x \pm \theta_\tau$ is nearly a multiple of 2π came under this heading). However, the last problem considered in §7.4, in which we took account of only the first term of (7.4.6) and assumed that K_x could be taken as constant, violated neither the paraxial approximation nor the periodicity of the reference planes; we may expect that this problem, and others of the same category, may be conveniently investigated by the matrix method of §7.2. As was implied in that section, the matrix method may be used for the analysis of certain perturbations if the accelerator is circular (or, rather, 'cyclic'), since the perturbation will not vitiate the fact that the 'dynamical transformation' represented by a complete revolution of the accelerator is exactly the same for every revolution. In order to emphasize that this and the succeeding sections of this chapter are particularly suitable for the analysis of circular accelerators (such as the strong-focusing bevatron), we shall, where appropriate, replace the characteristic exponent θ, which characterizes a section, by Θ which characterizes a complete revolution (cf. (7.2.26)).

Let us consider a perturbation of the variational function of the form

$$\Delta M = \eta x, \tag{7.5.1}$$

where $\eta(z)$ has the periodicity of the appropriate sequence of reference planes and it is assumed that any term in x' has been eliminated by the partial integration rule of §3.2. The perturbation characteristic function (7.4.1) may be expressed as

$$\Delta U_k = \alpha p_{xk} + \beta x_k; \qquad (7.5.2)$$

the coefficients are given by

$$\alpha = \int_{z_0}^{z_1} \eta h \, dz, \quad \beta = \int_{z_0}^{z_1} \eta g \, dz, \qquad (7.5.3)$$

if we assume that between reference planes (i.e. over any revolution of the accelerator) $x_k(z) = p_{xk} h(z) + x_k g(z)$. (7.5.4)

Now we see from (7.4.3) and (7.5.2) that the effective perturbations of p_{xk}, x_k are given by

$$\Delta p_{xk} = \beta, \quad \Delta x_k = -\alpha, \qquad (7.5.5)$$

so that the perturbation has the effect of replacing (7.2.1) by

$$\begin{pmatrix} p_{x,k+1} \\ x_{k+1} \end{pmatrix} = \begin{pmatrix} A & B \\ C & D \end{pmatrix} \begin{pmatrix} p_{xk} + \beta \\ x_k - \alpha \end{pmatrix}. \qquad (7.5.6)$$

The general solution of the inhomogeneous equation (7.5.6) may be expressed as the sum of any particular solution of this equation and the general solution of the homogeneous equation (7.2.1). It will therefore suffice to seek the steady solution of (7.5.6) which we write as p_x^S, x^S.

If we make use of (7.2.4) and (7.2.16), the steady solution of (7.5.6) may be written as

$$\begin{pmatrix} p_x^S \\ x^S \end{pmatrix} = \frac{1}{4 \sin^2 \frac{1}{2}\Theta} \begin{pmatrix} 1-A & -B \\ -C & 1-D \end{pmatrix} \begin{pmatrix} -\beta \\ \alpha \end{pmatrix}. \qquad (7.5.7)$$

We are assuming that α and β are small, and we now see from (7.5.7) that the perturbation will cause only slight displacement of the trajectories if Θ is not close to zero. Let us therefore consider the special case that $\Theta = \epsilon$, where ϵ is small. On deriving the appropriate approximations for A, B, C and D from (7.2.23), we find that

$$\left. \begin{aligned} p_x^S &= \epsilon^{-1}(\alpha \rho \operatorname{cosec} \psi + \beta \cot \psi), \\ x_S &= \epsilon^{-1}(\alpha \cot \psi + \beta \rho^{-1} \operatorname{cosec} \psi), \end{aligned} \right\} \qquad (7.5.8)$$

so that the perturbation will displace the trajectories appreciably if ϵ is smaller in magnitude than α and β.

Without going into details, we may see that the steady solution which we have obtained is consistent with the form (7.4.12) which referred to a problem of the same type as that we have just considered. It will be seen in §7.8 that perturbations of the type

(7.5.1) arise in the strong-focusing bevatron if the pole-pieces are subject to displacements from their ideal positions.

Another type of perturbation which we may expect to be tractable by matrix methods is that which produces a perturbation of the variational function of the form

$$\Delta M = \tfrac{1}{2}\eta x^2, \tag{7.5.9}$$

where $\eta(z)$ again has the same periodicity as the unperturbed system. A term of ΔM involving xx' could be absorbed into that shown; a term in x'^2 could if necessary be included on the right-hand side of (7.5.9). Perturbations such as (7.5.9) would arise if the focusing fields of an accelerator were perturbed in such a way as to affect the focusing power of the field without displacing the trajectory axis.

The perturbation characteristic function (7.4.1) may now be expressed as

$$\Delta U_k = \tfrac{1}{2}\alpha p_{xk}^2 + \beta p_{xk} x_k + \tfrac{1}{2}\gamma x_k^2, \tag{7.5.10}$$

where, from (7.5.4),

$$\alpha = \int_{z_0}^{z_1} \eta h^2 \,\mathrm{d}z, \quad \beta = \int_{z_0}^{z_1} \eta g h \,\mathrm{d}z, \quad \gamma = \int_{z_0}^{z_1} \eta g^2 \,\mathrm{d}z. \tag{7.5.11}$$

We see from (7.4.3) and (7.5.10) that the effective perturbations of p_{xk} and x_k are given by

$$\Delta p_{xk} = \beta p_{xk} + \gamma x_k, \quad x_k = -\alpha p_{xk} - \beta x_k. \tag{7.5.12}$$

If we again consider the case $\Theta = \epsilon$, where ϵ is small (the case $\Theta = \pm\pi \pm \epsilon$ may be considered in just the same way), the coefficients of the matrix appearing in (7.2.1) may be expanded as

$$\left.\begin{aligned} A &= 1 + \Delta A + \Delta^2 A + \dots, & B &= \Delta B + \Delta^2 \mathrm{B} + \dots,\\ C &= \Delta C + \Delta^2 C + \dots, & D &= 1 + \Delta D + \Delta^2 D + \dots, \end{aligned}\right\} \tag{7.5.13}$$

where, from (7.2.23) and (7.5.12),

$$\left.\begin{aligned} \Delta A &= \epsilon \cot\psi + \beta, & \Delta B &= -\rho\epsilon \operatorname{cosec}\psi + \gamma,\\ \Delta C &= \rho^{-1}\epsilon \operatorname{cosec}\psi + \alpha, & D &= -\epsilon\cot\psi - \beta; \end{aligned}\right\} \tag{7.5.14}$$

$\Delta^2 A$, etc., are of the second order in ϵ, α, β and γ, and so on.

The relation (7.2.16) shows that if $\Delta\Theta$ is the characteristic exponent for the *perturbed* system,

$$(\Delta\Theta)^2 - \tfrac{1}{12}(\Delta\Theta)^4 + \dots = -(\Delta A + \Delta D) - (\Delta^2 A + \Delta^2 D) - \dots. \tag{7.5.15}$$

However, the condition (7.2.4) imposes upon the coefficients appearing explicitly in (7.5.13) the restrictions

$$\Delta A + \Delta D = 0 \tag{7.5.16}$$

and

$$\Delta^2 A + \Delta^2 D = (\Delta A)^2 + \Delta B \Delta C. \tag{7.5.17}$$

Consequently (7.5.15) reduces, in lowest order, to

$$(\Delta\Theta)^2 = -(\Delta A)^2 - \Delta B \Delta C \tag{7.5.18}$$

which, it will be noted, involves only first-order terms.

Clearly the condition for stability is that the right-hand side of (7.5.18) should be positive; it if is negative, $\Delta\Theta$ is imaginary, so that the circular functions give way to hyperbolic functions. On substituting the values given by (7.5.14) into (7.5.18), we find that the latter may be rewritten as

$$(\Delta\Theta)^2 = (\epsilon - \epsilon_c)^2 - \epsilon_w^2, \tag{7.5.19}$$

where

$$\left.\begin{aligned} \epsilon_c &= \beta \cot\psi + \tfrac{1}{2}(\rho^{-1}\gamma - \rho\alpha)\operatorname{cosec}\psi, \\ \epsilon_w^2 &= \{\beta \operatorname{cosec}\psi + \tfrac{1}{2}(\rho^{-1}\gamma - \rho\alpha)\cot\psi\}^2 + \tfrac{1}{4}\{\rho^{-1}\gamma + \rho\alpha\}^2. \end{aligned}\right\} \tag{7.5.20}$$

We now see that there is stability as long as

$$|\epsilon - \epsilon_c| > \epsilon_w, \tag{7.5.21}$$

so that ϵ_c and ϵ_w denote the centre and half-width of a band of instability. We also see that the largest imaginary value taken by $\Delta\Theta$ is $i\epsilon_w$ at $\epsilon = \epsilon_c$, i.e. at the centre of the band. Hence the maximum exponential expansion of the beam corresponds to the exponent $k\epsilon_w$: the beam expands by a factor of e in ϵ_w^{-1} revolutions. The behaviour outside the band of instability is the same as that arising in the analogous problem treated in §6.5.

The second of equations (7.5.20) has an interesting consequence; any perturbation of the type (7.5.9) will result in potential instability unless it is *equivalent to* a perturbation in the value of Θ (ρ and ψ remaining fixed).

The reader will doubtless have observed an unsatisfactory feature of the above procedure; it was necessary to introduce the second-order terms of the coefficients of the perturbed matrix only to discover that these could be eliminated. It is true that this shortcoming was not serious in the example we have just considered, but it would lend unnecessary complication to a more difficult problem such as the coupling problem to be considered in §7.7.

There is another procedure, well known in quantum theory,‡ which does not possess this defect, and this will now be given.

We have seen that perturbations of the type (7.5.9) are liable to have disturbing effects upon the behaviour of a particle accelerator when the matrix representing a complete revolution is *degenerate*, that is when the characteristic roots are equal, being either 1 or -1; we shall find in §7.7 that the same observation holds even if we must consider two or more off-axis co-ordinates conjointly. Let us therefore consider a matrix M_{ij}, which in the present context need not be a 2×2 matrix, which has a degenerate root λ. This means that it is possible to find more than one linearly independent vectors $v_i^{(1)}, v_i^{(2)}, \ldots$, satisfying the homogeneous equation

$$(M_{ij}-\lambda\delta_{ij})\,v_j^{(r)}=0. \qquad (7.5.22)$$

There is another set of vectors $\tilde{v}_i^{(1)}, \tilde{v}_i^{(2)}, \ldots$, satisfying the transpose equation

$$\tilde{v}_i^{(r)}(M_{ij}-\lambda\delta_{ij})=0. \qquad (7.5.23)$$

If the matrix is now perturbed so that the typical element becomes $M_{ij}+\Delta M_{ij}$, we may expect the degenerate root λ to split into a number of different, but almost equal, roots. To each of the new roots will correspond a characteristic vector of the type v_i; a sufficiently flexible assumption is that this vector is expressible as a linear combination of the $v_i^{(r)}$ together with a term which will vanish when the perturbation vanishes. Hence we may assume that the perturbed matrix has a characteristic root $\lambda+\Delta\lambda$ to which there corresponds a characteristic vector $\alpha^{(r)}v_i^{(r)}+\Delta v_i$, where the $\alpha^{(r)}$ are scalar coefficients and the superfix r is summed over.

If we write down the perturbed form of (7.5.22) and then pick out the contribution which is of the first order in the perturbation, we obtain

$$(\Delta M_{ij}-\Delta\lambda\delta_{ij})\,\alpha^{(s)}v_j^{(s)}+(M_{ij}-\lambda\delta_{ij})\,\Delta v_j=0. \qquad (7.5.24)$$

If we now multiply this equation by $\tilde{v}_i^{(r)}$ and make use of (7.5.23), we obtain

$$\tilde{v}_i^{(r)}\,(\Delta M_{ij}-\Delta\lambda\delta_{ij})\,v_j^{(s)}\alpha^{(s)}=0, \qquad (7.5.25)$$

which is a homogeneous matrix equation for the coefficients $\alpha^{(s)}$. The perturbation of the characteristic root may therefore be found by writing down the condition that (7.5.25) has non-zero solutions,

‡ See, for instance, A. Landé, *Quantum Mechanics* (Pitman, London, 1951), pp. 84–8.

that is by equating the determinant of this matrix to zero:‡

$$\| \tilde{v}_i^{(r)}(\Delta M_{ij} - \Delta\lambda\delta_{ij}) v_j^{(s)} \| = 0. \qquad (7.5.26)$$

In order to avoid misunderstanding, this equation will be written out more fully for the special case of 'doublet' degeneracy (such as that considered in the preceding problem):

$$\begin{vmatrix} \tilde{v}_i^{(1)}\Delta M_{ij} v_j^{(1)} - \Delta\lambda \tilde{v}_i^{(1)} v_i^{(1)} & \tilde{v}_i^{(1)}\Delta M_{ij} v_j^{(2)} - \Delta\lambda \tilde{v}_i^{(1)} v_i^{(2)} \\ \tilde{v}_i^{(2)}\Delta M_{ij} v_j^{(1)} - \Delta\lambda \tilde{v}_i^{(2)} v_i^{(1)} & \tilde{v}_i^{(2)}\Delta M_{ij} v_j^{(2)} - \Delta\lambda \tilde{v}_i^{(2)} v_i^{(2)} \end{vmatrix} = 0. \qquad (7.5.27)$$

The suffixes i and j are of course summed over. This equation is quadratic in $\Delta\lambda$ as we should expect.

There is no difficulty in applying this method to the example which we have just considered in which the perturbed form of the matrix is given by (7.5.13). Since the unperturbed matrix is the unit matrix, the simplest choice of characteristic vectors is the following:

$$\begin{pmatrix} \tilde{v}_1^{(1)} \\ \tilde{v}_2^{(1)} \end{pmatrix} = \begin{pmatrix} v_1^{(1)} \\ v_2^{(1)} \end{pmatrix} = \begin{pmatrix} 1 \\ 0 \end{pmatrix}, \quad \begin{pmatrix} \tilde{v}_1^{(2)} \\ \tilde{v}_2^{(2)} \end{pmatrix} = \begin{pmatrix} v_1^{(2)} \\ v_2^{(2)} \end{pmatrix} = \begin{pmatrix} 0 \\ 1 \end{pmatrix}. \qquad (7.5.28)$$

The equation (7.5.27) therefore reduces to

$$\begin{vmatrix} \Delta A - \Delta\lambda & \Delta B \\ \Delta C & \Delta D - \Delta\lambda \end{vmatrix} = 0 \qquad (7.5.29)$$

which, on our noting that $\Delta\lambda = i\Delta\Theta$, is seen to be identical with (7.5.18). It is true that we have still to use the condition (7.5.16) between the coefficients of the perturbation of the matrix but, as we see from (7.5.14), this condition is automatically satisfied if the perturbation is derived from a perturbation characteristic function.

Although the latter method is in some respects an improvement on the former, it must be pointed out that, except in the special case in which the unperturbed matrix is diagonal, the characteristic vectors which appear in (7.5.27) will be complex. This, and the simplification of reducing the elements of (7.5.27) to simple terms as in (7.5.29), suggests that we attempt to replace the variables p_x and x by another pair of variables for which the matrix (7.2.1) becomes diagonal. Such a transformation will be given in the following section.

‡ This condition would of course simplify if, following the custom familiar in quantum mechanics, we specified that the vectors $\tilde{v}_i^{(r)}$ and $v_i^{(s)}$ should be normalized and orthogonal.

7.6. The diagonal representation‡

We have so far chosen to represent the behaviour of the projection of a trajectory upon the x-z plane either by the values of p_x and x in the reference planes or by the parameters K and κ, which are stationary in the paraxial approximation and in the absence of perturbations. Both these methods have merits, but it will be seen in this section that certain of the advantages of each appear in a third representation which replaces a pair of real variables by one complex variable.

Let us write the characteristic vector appearing in (7.2.7) as $(v, 1)$ so that, in the case of interest for which $\lambda = e^{i\theta}$,

$$\begin{pmatrix} A - e^{i\theta} & B \\ C & D - e^{i\theta} \end{pmatrix} \begin{pmatrix} v \\ 1 \end{pmatrix} = 0. \tag{7.6.1}$$

We find from (7.2.16), (7.2.20) and (7.2.21) that

$$v = \rho\, e^{i\psi}, \tag{7.6.2}$$

so that v is complex; should the section be symmetrical, so that $\psi = \frac{1}{2}\pi$, v is purely imaginary. The vectors $(v, 1)$ and $(\bar{v}, 1)$ are linearly independent, and it is therefore possible to express an arbitrary vector in the form

$$\begin{pmatrix} p_{x,k} \\ x_k \end{pmatrix} = \tfrac{1}{2} u_k \begin{pmatrix} v \\ 1 \end{pmatrix} + \tfrac{1}{2} \bar{u}_k \begin{pmatrix} \bar{v} \\ 1 \end{pmatrix}. \tag{7.6.3}$$

We note from this relation that x_k *is given by the real part of* u_k and $p_{x,k}$ by the real part of vu_k. Since the characteristic vector has the property (7.2.6), we infer at once that the u_k are related among themselves by

$$u_{k+1} = e^{i\theta} u_k. \tag{7.6.4}$$

This shows that the 'vector' u_k in the appropriate complex plane has constant magnitude and rotates about the origin; hence $|u_k|$ is an upper limit of $|x_k|$ so that we should expect to find that $|u_k| = K$, since K also is an upper limit of $|x_k|$.

We may take a different standpoint and say that the transformation (7.6.3) has diagonalized the matrix, since now

$$\begin{pmatrix} u_{k+1} \\ \bar{u}_{k+1} \end{pmatrix} = \begin{pmatrix} e^{i\theta} & \cdot \\ \cdot & e^{-i\theta} \end{pmatrix} \begin{pmatrix} u_k \\ \bar{u}_k \end{pmatrix}. \tag{7.6.5}$$

‡ This section is based upon unpublished work by J. S. Bell.

It is worth noticing, for possible application to perturbation problems, that if v were *not* given by (7.6.2), (7.6.5) would be replaced by a relation of the form

$$\begin{pmatrix} v_{k+1} \\ \bar{v}_{k+1} \end{pmatrix} = \begin{pmatrix} a & b \\ \bar{b} & \bar{a} \end{pmatrix} \begin{pmatrix} v_k \\ \bar{v}_k \end{pmatrix}, \quad (7.6.6)$$

where, as a consequence of (7.2.4) and the properties of Jacobians,

$$\bar{a}a - \bar{b}b = 1. \quad (7.6.7)$$

The variable u_k may be expressed in terms of $p_{x,k}$ and x_k as

$$u_k = -\mathrm{i}\sigma^{-1}p_{x,k} + (1 + \mathrm{i}\cot\psi)\, x_k, \quad (7.6.8)$$

but its relation to K and κ takes the simpler form

$$u_k = -\mathrm{i}K\,\mathrm{e}^{\mathrm{i}(k\theta+\kappa)}. \quad (7.6.9)$$

We note in particular that

$$K = |\, u_k \,| \quad (7.6.10)$$

as we had anticipated. In consequence, the formula (7.3.5) for the adiabatic invariant may be rewritten as

$$\mathsf{A} = \pi\sigma\, |\, u_k \,|^2. \quad (7.6.11)$$

In order to evaluate perturbation characteristic functions, it will be necessary to express $x_k(z)$, i.e. the value of x between the kth and $(k+1)$th reference planes, as a function of u_k and $\bar{u}_k$. If we write

$$x_k(z) = \bar{u}_k f(z) + u_k \bar{f}(z), \quad (7.6.12)$$

we readily find from (7.6.3) and (7.5.4) that $f(z)$ is related to $g(z)$ and $h(z)$ by

$$f(z) = \tfrac{1}{2}(g(z) + \bar{v}h(z)). \quad (7.6.13)$$

If the perturbation differential relation (7.4.2) is specialized to involve only p_x and x, and if we then transform to the diagonal representation, we obtain

$$\delta\Delta U_k = (v - \bar{v})(\Delta u_k \delta\bar{u}_k - \Delta\bar{u}_k \delta u_k) \quad (k \text{ n.t.b.s.}), \quad (7.6.14)$$

from which it follows that the effective perturbation of the complex variable u_k is given by

$$\Delta u_k = -\tfrac{1}{2}\mathrm{i}\sigma^{-1}\frac{\partial \Delta U_k}{\partial \bar{u}_k} \quad (k \text{ n.t.b.s.}). \quad (7.6.15)$$

In arriving at (7.6.15) we have used the relation

$$v - \bar{v} = 2\mathrm{i}\sigma \quad (7.6.16)$$

which follows from (7.6.2) and (7.2.24). Should it be necessary to consider two or more off-axis co-ordinates simultaneously, it would be necessary to introduce a corresponding number of complex variables; we should then obtain a set of equations of the form (7.6.15).

It is interesting to repeat, as briefly as possible, the two perturbation calculations which were carried out in the preceding section in terms of the diagonal representation. If the perturbation of the variational function is of the form (7.5.1), the corresponding characteristic function is found to be

$$\Delta U_k = \alpha \bar{u}_k + \bar{\alpha} u_k, \tag{7.6.17}$$

where

$$\alpha = \int_{z_0}^{z_1} \eta f \, dz. \tag{7.6.18}$$

If $\Theta = \epsilon$, where ϵ is small, (7.6.4) must be replaced by

$$u_{k+1} = (1 + i\epsilon) u_k + \Delta u_k \tag{7.6.19}$$

where, from (7.6.15) and (7.6.17),

$$\Delta u_k = -\tfrac{1}{2}\sigma^{-1}\alpha. \tag{7.6.20}$$

The steady solution of (7.6.19) is

$$u_k^S = \tfrac{1}{2}\epsilon^{-1}\sigma^{-1}\alpha \tag{7.6.21}$$

so that K^S, which determines the maximum displacement of the steady trajectory, is given by

$$K^S = \tfrac{1}{2}\epsilon^{-1}\sigma^{-1}|\alpha|. \tag{7.6.22}$$

If the perturbation of the variational function is of the form (7.5.9), the perturbation characteristic function is expressible as

$$U_k = \tfrac{1}{2}\alpha \bar{u}_k^2 + \beta \bar{u}_k u_k + \tfrac{1}{2}\bar{\alpha} u_k^2 \quad (k \text{ n.t.b.s.}) \tag{7.6.23}$$

where α, which is complex, and β, which is real, are given by

$$\alpha = \int_{z_0}^{z_1} \eta f^2 \, dz, \quad \beta = \int_{z_0}^{z_1} \eta \bar{f} f \, dz. \tag{7.6.24}$$

If we again suppose that $\Theta = \epsilon$ and if we use (7.6.15), we see that the diagonal matrix relation (7.6.5) is now replaced by a relation of the form (7.6.6), where

$$a = 1 + i\epsilon + i\Delta a + \ldots, \quad b = \Delta b + \ldots; \tag{7.6.25}$$

Δa, which is real, and Δb are related to α and β by

$$\Delta a = -\tfrac{1}{2}\sigma^{-1}\beta, \quad \Delta b = -\tfrac{1}{2}i\sigma^{-1}\alpha. \tag{7.6.26}$$

Since the unperturbed matrix is diagonal, the equation (7.5.27) for the characteristic roots will again reduce to the form (7.5.29).

If we write $\Delta\lambda = i\Delta\Theta$ and find the perturbation of the matrix in (7.6.6) from (7.6.25), we obtain in place of (7.5.29)

$$\begin{vmatrix} i\epsilon + i\Delta a - i\Delta\Theta & \Delta b \\ \Delta\bar{b} & -i\epsilon - i\Delta a - i\Delta\Theta \end{vmatrix} = 0 \qquad (7.6.27)$$

or

$$(\Delta\Theta)^2 = (\epsilon + \Delta a)^2 - |\Delta b|^2. \qquad (7.6.28)$$

We see, as before, that the accelerator remains stable in the presence of the perturbation only if Θ lies outside a band of instability centred at $\Theta = -\Delta a$ and of half-width $|\Delta b|$; there will be a similar band of instability near to $\Theta = \pm\pi$. These expressions must be acknowledged to be simpler than those of (7.5.20) but it should be remembered that the integrals (7.6.24) involve the complex function $f(z)$ rather than the real functions $g(z)$ and $h(z)$.

The formulae which we have established in this section will be applied in §7.8 to the problem of determining tolerances for certain constructional errors in the strong-focusing bevatron; for this purpose we shall need to *estimate* integrals such as (7.6.18) and (7.6.24) rather than calculate them exactly. We have already seen that it is the absolute value of such integrals which is required, not their real and imaginary parts; if we replace f by its modulus $|f|$, the integrals will be real and so simpler to evaluate and we may, moreover, expect estimates obtained in this way to lead to slightly stricter, rather than laxer, tolerances. Let us therefore investigate how $|f|$ varies in the $\frac{1}{2}\pi$ mode which we considered in §7.2 and which we shall adopt in §7.8.

We find from (7.5.4), (7.6.2) and (7.6.13) that

$$|f|^2 = \tfrac{1}{4}(g^2 + 2\rho\cos\psi\, gh + \rho^2 h^2). \qquad (7.6.29)$$

Equations (7.2.32) shows that in the range $0 < z < L$,

$$g(z) = \cos\omega z, \quad h(z) = \rho^{-1}\sin\omega z, \qquad (7.6.30)$$

so that

$$|f|^2 = \tfrac{1}{4}(1 + \cos\psi \sin 2\omega z). \qquad (7.6.31)$$

Similarly we find from (7.2.32) and (7.2.33) that in the range $L < z < 2L$,

$$g(z) = -\sinh\omega z, \quad h(z) = \rho^{-1}\cosh\omega z, \qquad (7.6.32)$$

so that

$$|f|^2 = \tfrac{1}{4}\operatorname{sech}\tfrac{1}{2}\pi \cosh(2\omega(z - \tfrac{3}{2}L)). \qquad (7.6.33)$$

It may be verified that the variation of $|f|^2$ in the succeeding convergent and divergent fields is identical with (7.6.31) and (7.6.33); since θ has changed by π over these two complete sections, we may conclude that $|f|^2$ behaves as (7.6.31) in every convergent field and as (7.6.33) in every divergent field.

The function defined by (7.6.31) and (7.6.33) is shown in fig. 7.5, and it is seen that $|f|$ remains within 40 % of the value which it takes in the reference planes. If the order of converging and diverging sections is reversed, we find (noting that the sign of $\cos\psi$ is changed) that (7.6.31) and (7.6.33) are interchanged. We may therefore infer from fig. 7.5 that, for *rough* estimates, we may give $|f|$ a constant value.

It is interesting to compare the slight variation of $|f|$ shown in fig. 7.5 with the large variation of g and h which we must expect

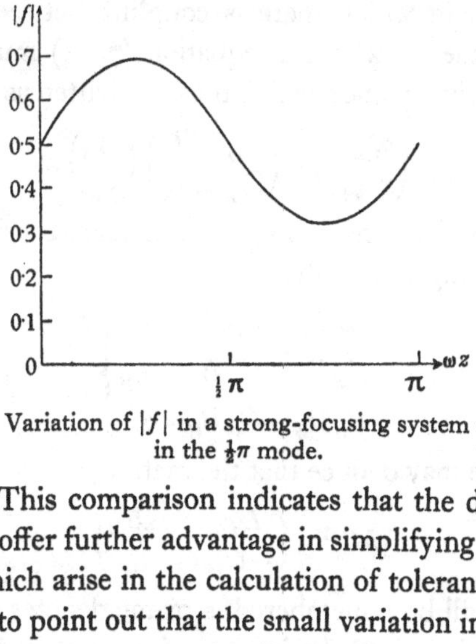

Fig. 7.5. Variation of $|f|$ in a strong-focusing system operating in the $\frac{1}{2}\pi$ mode.

from fig. 7.3. This comparison indicates that the diagonal representation may offer further advantage in simplifying the estimation of integrals which arise in the calculation of tolerances. However, it is necessary to point out that the small variation in $|f|$ which we have found in a system operating in the $\frac{1}{2}\pi$ mode holds only for the choice of reference planes first considered in § 7.2; if we change to either of the symmetrical representations indicated in (7.2.38) and (7.2.40), we find that the variation of $|f|$ is comparable with that of g and h.

7.7. Instability due to linear coupling

The power of the diagonal representation is well illustrated by the ease with which one can investigate, by this method, the effect of a coupling term in the paraxial approximation to the variational function, for the solution of this problem by either the method of variation of parameters or the matrix method of §7.5 with the representation of §7.2 is quite laborious. We shall see in the next section that such a coupling term arises in the strong-focusing bevatron if consecutive magnet sectors are misalined in such a way as to appear to be twisted about the trajectory axis.

Before proceeding with this particular problem, it is expedient to establish the generalization of the rule which we noted in §7.2 for the particular case of a paraxial focusing system involving only one off-axis co-ordinate, namely, that *if λ is a characteristic root of the matrix describing a linear periodic dynamical system, so is λ^{-1}.* This theorem is associated with the name of Poincaré.‡

In a system in which there is coupling between two off-axis co-ordinates, the 2×2 matrix equation (7.2.1) must be replaced by a 4×4 matrix equation which may be written as

$$\begin{pmatrix} p_{i,k+1} \\ x_{i,k+1} \end{pmatrix} = \begin{pmatrix} A_{ij} & B_{ij} \\ C_{ij} & D_{ij} \end{pmatrix} \begin{pmatrix} p_{j,k} \\ x_{j,k} \end{pmatrix}. \tag{7.7.1}$$

We may verify, by considering the Lagrange brackets $\{p_{i,k}, x_{j,k}\}$, $\{p_{i,k}, x_{j,k}\}$ and $\{x_{i,k}, x_{j,k}\}$, that

$$\left.\begin{aligned} A_{li}C_{lj} - C_{li}A_{lj} &= 0, \\ A_{li}D_{lj} - C_{li}B_{lj} &= \delta_{ij}, \\ B_{li}D_{lj} - D_{li}B_{lj} &= 0, \end{aligned}\right\} \tag{7.7.2}$$

from which we may deduce that the matrix

$$M^{-1} = \begin{pmatrix} D'_{ij} & -B'_{ij} \\ -C'_{ij} & A'_{ij} \end{pmatrix} \tag{7.7.3}$$

(in which, it will be remembered, a prime denotes a transpose, so that $A'_{ij} = A_{ji}$) is indeed the inverse of the matrix M appearing in (7.7.1). However, the relation between M and its inverse may be written as

$$M^{-1} = J^{-1}M'J, \tag{7.7.4}$$

where

$$J = \begin{pmatrix} \cdot & -\delta_{ij} \\ \delta_{ij} & \cdot \end{pmatrix}. \tag{7.7.5}$$

It is a consequence of the form of the relation (7.7.4) that if λ is a characteristic root of M so that

$$\|M - \lambda I\| = 0 \tag{7.7.6}$$

‡ E. T. Whittaker, *Analytical Dynamics* (Cambridge University Press, 3rd ed. 1927), pp. 403, 404.

(where I is the unit matrix), λ is also a root of M^{-1} or, equivalently, λ^{-1} is a root of M. This theorem holds, of course, for any number of co-ordinates.

In the case in which we are at present interested, in which there are only two co-ordinates, the above theorem shows that the characteristic equation must be of the form

$$\lambda^4+s\lambda^3+t\lambda^2+s\lambda+1=0. \tag{7.7.7}$$

We now know that if λ is a root, so also are $\bar{\lambda}$, λ^{-1} and $\bar{\lambda}^{-1}$. Hence the roots may either fall into two pairs, each pair being of the form $e^{\pm i\theta}$, $e^{\pm\mu}$ or $-e^{\pm\mu}$, or they may form the group $e^{\mu+i\theta}$, $e^{\mu-i\theta}$, $e^{-\mu+i\theta}$ and $e^{-\mu-i\theta}$. Clearly the transition under perturbation from characteristic exponents $\pm i\theta$ to $\pm\mu$ can occur only if θ is close to 0 or π, but the transition under perturbation from the pairs $\pm i\theta_x$, $\pm i\theta_y$ to the set $\pm\mu\pm i\theta$ can occur if $\theta_x\approx\theta_y$ or if $\theta_x\approx-\theta_y$. These two cases will call for separate investigation.

We take the unperturbed form of the variational function to be the sum of two terms of the type (7.2.31), one in x and the other in y; we shall find that a field sector which is convergent for x is divergent for y and vice versa. The perturbation of M will be taken to be of the form

$$\Delta M=\eta xy. \tag{7.7.8}$$

It is worth pointing out that a magnetic field with a component along the trajectory axis would introduce a term in $x'y-xy'$ which could be taken into account in (7.7.8).

If there is paraxial coupling, the perturbation characteristic function is expressible as

$$\Delta U=\alpha\bar{u}_x u_y+\bar{\alpha}u_x\bar{u}_y+\beta\bar{u}_x\bar{u}_y+\bar{\beta}u_x u_y. \tag{7.7.9}$$

(Since it is necessary to attach suffixes x and y to u, we shall suppress the suffix k.) The coefficients in (7.7.9) are given by

$$\alpha=\int_{z_0}^{z_1}\eta f_x\bar{f}_y\,dz,\quad \beta=\int_{z_0}^{z_1}\eta f_x f_y\,dz. \tag{7.7.10}$$

On generalizing (7.6.15) to include u_x and u_y, we find that the effective perturbations of u_x and u_y are given by

$$\Delta u_x=-\tfrac{1}{2}i\sigma_x^{-1}(\alpha u_y+\beta\bar{u}_y),\quad \Delta u_y=-\tfrac{1}{2}i\sigma_y^{-1}(\bar{\alpha}u_x+\beta\bar{u}_x). \tag{7.7.11}$$

Let us now investigate the possibility of instability when $\Theta_x \approx \Theta_y$ by writing

$$\Theta_x = \Theta, \quad \Theta_y = \Theta + \epsilon, \tag{7.7.12}$$

where ϵ is small, but impose the restriction that Θ is not close to zero or $\pm\pi$. (Cases such as the latter, which involve higher degeneracies, could nevertheless be studied by the same method.) It is proposed that we regard as the unperturbed system that in which there is no coupling of the form (7.7.8) and $\epsilon = 0$. Then the matrix relation describing one revolution of the accelerator is

$$\begin{pmatrix} u_{x,k+1} \\ \bar{u}_{x,k+1} \\ u_{y,k+1} \\ \bar{u}_{y,k+1} \end{pmatrix} = \begin{pmatrix} e^{i\Theta} & \cdot & \cdot & \cdot \\ \cdot & e^{-i\Theta} & \cdot & \cdot \\ \cdot & \cdot & e^{i\Theta} & \cdot \\ \cdot & \cdot & \cdot & e^{-i\Theta} \end{pmatrix} \begin{pmatrix} u_{x,k} \\ \bar{u}_{x,k} \\ u_{y,k} \\ \bar{u}_{y,k} \end{pmatrix}. \tag{7.7.13}$$

The perturbation of this matrix, which represents both the effect of coupling and the difference between Θ_x and Θ_y, is given by

$$\Delta M = \begin{pmatrix} \cdot & \cdot & -\tfrac{1}{2}i\sigma_x^{-1}e^{i\Theta}\alpha & -\tfrac{1}{2}i\sigma_x^{-1}e^{i\Theta}\beta \\ \cdot & \cdot & \tfrac{1}{2}i\sigma_x^{-1}e^{-i\Theta}\bar{\beta} & \tfrac{1}{2}i\sigma_x^{-1}e^{-i\Theta}\bar{\alpha} \\ -\tfrac{1}{2}i\sigma_y^{-1}e^{i\Theta}\bar{\alpha} & -\tfrac{1}{2}i\sigma_y^{-1}e^{i\Theta}\beta & i\epsilon e^{i\Theta} & \cdot \\ \tfrac{1}{2}i\sigma_y^{-1}e^{-i\Theta}\bar{\beta} & \tfrac{1}{2}i\sigma_y^{-1}e^{-i\Theta}\alpha & \cdot & -i\epsilon e^{-i\Theta} \end{pmatrix}. \tag{7.7.14}$$

We may adopt as the characteristic vectors corresponding to the root $e^{i\Theta}$ of the unperturbed matrix

$$(\tilde{v}_i^{(1)}) = (v_i^{(1)}) = \begin{pmatrix} 1 \\ \cdot \\ \cdot \\ \cdot \end{pmatrix}, \quad (\tilde{v}_i^{(2)}) = (v_i^{(2)}) = \begin{pmatrix} \cdot \\ \cdot \\ 1 \\ \cdot \end{pmatrix}. \tag{7.7.15}$$

If we now suppose that the characteristic exponent of the perturbed matrix which is near to Θ is $\Theta + \Delta\Theta$, the perturbation of the characteristic root is $i\Delta\Theta\, e^{i\Theta}$. The characteristic equation (7.5.27) now becomes

$$\begin{vmatrix} -i\Delta\Theta\, e^{i\Theta} & -\tfrac{1}{2}i\sigma_x^{-1}e^{i\Theta}\alpha \\ -\tfrac{1}{2}i\sigma_y^{-1}e^{i\Theta}\bar{\alpha} & i\epsilon e^{i\Theta} - i\Delta\Theta\, e^{i\Theta} \end{vmatrix} = 0, \tag{7.7.16}$$

i.e.

$$(\Delta\Theta - \tfrac{1}{2}\epsilon)^2 = \tfrac{1}{4}\epsilon^2 + \tfrac{1}{4}\sigma_x^{-1}\sigma_y^{-1}|\alpha|^2. \tag{7.7.17}$$

Now according to the conventions adopted in §7.2 in defining θ, etc., σ is always positive; $\Delta\Theta$ as defined by (7.7.17) is therefore always real, so that in this case there is no instability.

Let us now proceed to investigate the other possible instability at $\Theta_x \approx -\Theta_y$. There is no difficulty in repeating the preceding calculations but we may see what the result will be by noting from (7.2.23) that, on temporarily relaxing our earlier conventions, the representation of a system by the coefficients θ, ψ and ρ is unchanged if the signs of θ and ψ are changed simultaneously; this would entail a reversal of the sign of σ. Hence the results for the case

$$\Theta_x = -\Theta, \quad \Theta_y = \Theta + \epsilon \tag{7.7.18}$$

may be deduced from (7.7.17) by reversing the sign of σ_x and noting that since u_x and $\bar{u}_x$ are now interchanged, α and $\bar{\beta}$ also must be interchanged. In this way we obtain

$$(\Delta\Theta - \tfrac{1}{2}\epsilon)^2 = \tfrac{1}{4}\epsilon^2 - \tfrac{1}{4}\sigma_x^{-1}\sigma_y^{-1}|\beta|^2. \tag{7.7.19}$$

We now see that, since σ_x and σ_y as they appear in (7.7.19) are both positive, there is a band of instability centred at $\Theta_x = -\Theta_y$ and of half-width ϵ_w, where

$$\epsilon_w = (\sigma_x \sigma_y)^{-\frac{1}{2}} |\beta|. \tag{7.7.20}$$

We may also assert that the maximum rate of expansion of the beam (at the centre of the band) is exponential with exponent $k\epsilon_w$, and we may infer from the investigation of § 6.5 that immediately outside the band of instability the x and y amplitudes oscillate between large and small values.

It is rather surprising to find in the analysis of this problem that, in the presence of coupling, one degeneracy leads to instability but the other does not, and it is instructive to inquire why this should be so. The following explanation is due to J. S. Bell.

We have already seen in § 7.3 that it is possible to form an invariant (7.3.7) characterizing a paraxial focusing system involving only one off-axis co-ordinate by forming a Lagrange invariant (7.3.8) from an arbitrary trajectory and the displacement of this trajectory through one section. There is no reason why the same procedure should not be applied to a focusing system involving both x and y; if, in the first instance, we ignore the possibility of coupling between the two co-ordinates, we see that the resulting Lagrange invariant is

$$\begin{aligned} \mathsf{L} = (C_x p_x^2 - (A_x - D_x) p_x x - B_x x^2) \\ + (C_y p_y^2 - (A_y - D_y) p_y y - B_y y^2). \end{aligned} \tag{7.7.21}$$

The argument linking (7.3.9) to (7.3.7) may now be repeated. The accelerator will be convergent if the surface $\mathsf{L}=$const. drawn in (p_x, x, p_y, y) space is closed, that is, if the quadratic form (7.7.21) is of definite sign. This property is ensured if each of the quadratic forms in p_x, x and p_y, y are definite and of the same sign. We have already seen that the former condition is equivalent to (7.2.14).

Let us now consider the possibility of paraxial coupling terms between x and y. The effect of such a perturbation upon the quadratic form (7.7.21) is to change slightly the values of the coefficients A_x, etc., and to introduce 'mixed' terms of the form $p_x p_y$, etc., which, however, have small coefficients. If we now specify that θ_x and θ_y should be real and should not be close to zero or $\pm\pi$, we see that *the quadratic form* (7.7.21) *will be positive-definite provided that C_x and C_y have the same sign. Moreover, if these conditions are satisfied the quadratic form will remain positive-definite even if a coupling perturbation introduces other terms with small coefficients.*

It is easy to see that the above result is in agreement with our earlier conclusions, for C_x and C_y have the same signs if $\theta_x \approx \theta_y$ but opposite signs if $\theta_x \approx -\theta_y$. It must be emphasized, however, that the second demonstration of the stability of one of the potential instabilities is much stronger than the first in that it is not restricted to first-order perturbation theory.‡ In general, Bell's method offers a simple and reliable method of shortening the list of potential instabilities of a machine which one might draw up by general considerations based upon Poincaré's theorem.

7.8. The strong-focusing synchrotron

In the earlier sections of this chapter we have already considered what perturbations we must expect to arise in circular periodic-focusing accelerators and under what conditions such perturbations will give rise to instability. It is proposed that in this section we investigate how the preceding theory may be used to estimate certain tolerances for the construction of a strong-focusing synchrotron. By confining our attention to the betatron oscillations, i.e. by

‡ One can, however, assert that if there are two real and different values for Θ to first order, there cannot be imaginary contributions to Θ to higher order, since this would imply the existence of more than four characteristic roots. Since this provision is generally satisfied, the statements of first-order perturbation theory are stronger than one might at first imagine.

ignoring the synchrotron oscillations, we effectively simplify our problem to one of static electron optics.

Let us adopt the same axes as in our treatment of the synchrotron in §6.4; the trajectory axis is a circle of radius a, the x-axis is directed radially inwards and the y-axis completes the orthogonal set (x, y, z). Since we are proceeding by static electron optics, we must relace the variational function M by the variational function m. The focusing field in the strong-focusing synchrotron is a magnetic field of mirror symmetry; the appropriate variational function may therefore be derived from §5.2. If the strength of the magnetic field upon the trajectory axis is related to the particle momentum by

$$\mathsf{H}_y = -a^{-1}\mathsf{p}, \tag{7.8.1}$$

$m^{(1)}$, as given by (5.2.14), will vanish. (If the synchrotron is to accelerate protons, it is necessary to change the sign attached to $\mathbf{H}$ in (5.2.14) and (5.2.15).) The function $m^{(2)}$ given by (5.2.15) now reduces to

$$m^{(2)} = \tfrac{1}{2}\mathsf{p}(x'^2+y'^2) + \tfrac{1}{2}(\mathsf{p}a^{-2} - \mathsf{H}_{y,x})\,x^2 + \tfrac{1}{2}\mathsf{H}_{y,x}y^2. \tag{7.8.2}$$

Let us now assume that the magnetic field is divided into $2N$ sectors of equal length L, so that

$$L = \pi N^{-1} a; \tag{7.8.3}$$

the gradient of the field is to alternate from sector to sector, and each section, composed of two adjacent sectors, is to operate in approximately the $\tfrac{1}{2}\pi$ mode. If, in the section extending from $z=0$ to $z=2L$, we suppose that

$$\left.\begin{aligned} H_{y,x} &= -(n+\Delta n_1)\,a^{-1}\mathsf{H}_y \quad (0<z<L),\\ &= \ (n+\Delta n_2)\,a^{-1}\mathsf{H}_y \quad (L<z<2L),\end{aligned}\right\} \tag{7.8.4}$$

we may conveniently divide (7.8.2) into two parts, writing

$$\left.\begin{aligned} m^{(2)} &= \tfrac{1}{2}\mathsf{p}(x'^2+y'^2) - \tfrac{1}{2}\mathsf{p}a^{-2}n(x^2-y^2) \quad (0<z<L),\\ &= \tfrac{1}{2}\mathsf{p}(x'^2+y'^2) + \tfrac{1}{2}\mathsf{p}a^{-2}n(x^2-y^2) \quad (L<z<2L),\end{aligned}\right\} \tag{7.8.5}$$

and

$$\left.\begin{aligned} \Delta m^{(2)} &= -\tfrac{1}{2}\mathsf{p}a^{-2}(\Delta n_1 - 1)\,x^2 + \tfrac{1}{2}\mathsf{p}a^{-2}\Delta n_1 y^2 \quad (0<z<L),\\ &= \ \tfrac{1}{2}\mathsf{p}a^{-2}(\Delta n_2 + 1)\,x^2 - \tfrac{1}{2}\mathsf{p}a^{-2}\Delta n_2 y^2 \quad (L<z<2L).\end{aligned}\right\} \tag{7.8.6}$$

The terms of (7.8.5) involving x are now of the form (7.2.31), where

$$\omega = n^{\frac{1}{2}}a^{-1}; \tag{7.8.7}$$

the terms involving y are of the same form except that the convergent and divergent sectors are interchanged.

If we suppose that $n \gg 1$ and that Δn_1 and Δn_2 are comparable with unity, we may regard (7.8.6) as a perturbation of the variational function (7.8.5). If the accelerator is to operate in the $\frac{1}{2}\pi$ mode according to the approximation involved in replacing (7.8.2) by (7.8.5), we must have $\omega L = \frac{1}{2}\pi$ or, equivalently,

$$n = \tfrac{1}{4}N^2. \tag{7.8.8}$$

Let us now estimate the departure of the characteristic exponent θ from $\frac{1}{2}\pi$ due to the terms collected in (7.8.6). By integrating the terms of (7.8.6) involving x over a complete section, we obtain a characteristic function

$$\Delta U = \tfrac{1}{2}\alpha p_x^2 + \beta p_x x + \tfrac{1}{2}\gamma x^2 \tag{7.8.9}$$

where, from (7.5.4) and (7.8.5),

$$\alpha = -\mathsf{p}a^{-2}(\Delta n_1 - 1)\int_0^L h^2\,dz + \mathsf{p}a^{-2}(\Delta n_2 + 1)\int_L^{2L} h^2\,dz, \quad \text{etc.}; \tag{7.8.10}$$

$g(z)$ and $h(z)$ are given by (7.6.30) and (7.6.32). We find from (7.4.3) that the effective perturbations of p_x and x are given by

$$\Delta p_x = \beta p_x + \gamma x, \quad \Delta x = -\alpha p_x - \beta x. \tag{7.8.11}$$

(The suffix k has now been suppressed.) This may be represented as a perturbation of the elements of (7.2.1) of amount

$$\left.\begin{aligned} \Delta A &= A\beta - B\alpha, & \Delta B &= A\gamma - B\beta, \\ \Delta C &= C\beta - D\alpha, & \Delta D &= C\gamma - D\beta. \end{aligned}\right\} \tag{7.8.12}$$

If the characteristic exponent of the perturbed matrix is to be $\frac{1}{2}\pi + \Delta\theta_x$, we see from (7.2.16) that

$$\Delta\theta_x = -(A-D)\,\beta + B\alpha - C\gamma; \tag{7.8.13}$$

the appropriate values of A, B, C, D are given by (7.2.36). On evaluating the integrals and remembering the relation (7.2.26), we arrive at the result

$$\Delta\Theta_x = N^{-1}\{(2\sinh\tfrac{1}{2}\pi + \pi\cosh\tfrac{1}{2}\pi)(\Delta n_1 - 1) - 2\sinh\tfrac{1}{2}\pi(\Delta n_2 + 1)\}. \tag{7.8.14}$$

Similarly, or by inspection, we may establish the formula

$$\Delta\Theta_y = N^{-1}\{-2\sinh\tfrac{1}{2}\pi\,.\,\Delta n_1 + (2\sinh\tfrac{1}{2}\pi + \pi\cosh\tfrac{1}{2}\pi)\,\Delta n_2\}. \tag{7.8.15}$$

Let us suppose, as a numerical example, that $N = 40$ so that there are 80 sectors altogether; then $n = 400$. Equation (7.8.14) shows that even if $\Delta n_1 = \Delta n_2 = 0$ the term due to the curvature of the trajectory axis produces a shift of 0·1 radian or about 6° in Θ_x. Hence it is in theory possible, by judicious choice of Δn_1 and Δn_2, to change Θ_x and Θ_y substantially, although individual sections will remain effectively unchanged. There will therefore be no inconsistency when in subsequent perturbation calculations we assume that Θ_x and Θ_y are to some extent variable but also that individual sections operate in the $\frac{1}{2}\pi$ mode. In practice one would find it very difficult to control the gradient coefficient n to within fine limits, but it is nevertheless essential to find *some* method of ensuring that the characteristic exponents avoid the critical values and relations which we have already found to exist.

It is seen from the paragraph following (7.3.6) that, in the present problem, the adiabatic variation of the amplitude of the betatron oscillations is proportional to $\mathsf{p}^{-\frac{1}{2}}$. If the injection energy is 10 MeV. and the extraction energy 10,000 MeV., the betatron amplitude contracts by a factor of almost ten. This contraction must, of course, be set against expansion of the beam due to gas scattering.

Let us now attempt to assign a limit to the permissible displacement of pole-pieces in the x- and y-directions. If a sector is displaced in the x-direction a distance ξ, the perturbed form of the variational function may be obtained by replacing x in (7.8.5) by $x-\xi$. The perturbation of m is therefore of the form (7.5.1), where

$$\eta = \pm \mathsf{p} a^{-2} n \xi. \tag{7.8.16}$$

It was suggested in §7.6 that we might estimate the absolute value of integrals such as (7.6.18) by replacing f by its average absolute value $\frac{1}{2}$; hence we may adopt as the absolute value of (7.6.18) for any one sector $\frac{1}{2}|\eta|L$.

To different types of constructional error will correspond different assumptions as to the statistical distribution of the variable ξ among sectors. It is proposed that we here assume that there is no correlation between the values taken by ξ in different sectors. If we now note that the root-mean-square (r.m.s.) of a sum of n independent similar variables, whose mean values are zero, is $n^{\frac{1}{2}}$ times the r.m.s. of a single variable,‡ we may estimate that the

‡ We may also assert, on the basis of the central limit theorem (H. Cramér, *Random Variables and Probability Distributions* (Cambridge University Press, 1937), p. 109) that with these and certain other rather weak assumptions, the distribution is approximately normal if n is large.

absolute value of the sum of the integrals (7.6.18) over the $2N$ sectors has the expectation value

$$|\alpha|_{\text{exp.}} = (2N)^{\frac{1}{2}}\, \tfrac{1}{2} L \mathsf{p} a^{-2} n \xi_{\text{r.m.s.}}. \tag{7.8.17}$$

Hence, from (7.6.22) and (7.2.37),

$$K^S_{\text{exp.}} \approx 1{\cdot}4 N^{\frac{1}{2}} \epsilon^{-1} \xi_{\text{r.m.s.}}. \tag{7.8.18}$$

This formula, which effectively represents the deviation from the trajectory axis to be expected from the misalinement, holds for both the x- and y-directions.

If we specify, in the case we have already considered, that $N = 40$, that K is not to exceed 1 cm. even if ϵ is as small as 0·05 radian (i.e. if Θ is as close to zero as $3°$), we obtain a tolerance on ξ of 0·006 cm., i.e. of 60 μ.

The type of constructional error which we have just considered may be regarded as one which affects H_y, the field strength upon the trajectory axis (and also H_x, although we have not considered this explicitly), but leaves $\mathsf{H}_{y,x}$, the gradient of the field strength, unchanged. Let us now consider a constructional error which affects the field gradient; it is most convenient to obtain the appropriate tolerance in terms of n defined by (7.8.4). Once again it will suffice to consider the co-ordinate x; we see from (7.8.5) that if n is changed by amount Δn, the perturbation of m is of the form (7.5.9), where

$$\eta = \pm \mathsf{p} a^{-2} \Delta n. \tag{7.8.19}$$

We saw in §§7.5 and 7.6 that perturbations of this type produce bands of instability near $\Theta = 0$ and $\pm\pi$. In order to determine the width of these bands, we must evaluate the integrals (7.6.24). If we again assume that there is no correlation between the values of Δn in different sectors, we may use the same statistical method as before to estimate the absolute value of the integrals (7.6.24) and hence of the terms (7.6.26). In this way we obtain

$$\epsilon_{\text{exp.}} \approx 0{\cdot}7 N^{\frac{1}{2}} \frac{(\Delta n)_{\text{r.m.s.}}}{n} \tag{7.8.20}$$

for the expectation value of the half-width of the band of instability; the expectation value of the displacement of the centre of the band from 0 or $\pm\pi$ is given by the same formula. Hence if we stipulate that even if Θ approaches to within ϵ of 0 or $\pm\pi$ the machine will nevertheless be operating one half-width away from the edge of the band of instability (so that, from the results of § 6.5, the amplitude of oscillation varies by no more than a factor of $\sqrt{3}$), the r.m.s. error in n must satisfy

$$\frac{(\Delta n)_{\text{r.m.s.}}}{n} < 0\cdot 5 N^{-\frac{1}{2}} \epsilon. \qquad (7.8.21)$$

If once again $N=40$ and $\epsilon = 0\cdot 05$ (3°), we see that the r.m.s. error in n must be less than about $\frac{1}{2}$ %.

Now let us obtain a tolerance for 'twist' misalinements. By replacing x and y in (7.8.5) by $x+\zeta y$ and $y-\zeta x$, respectively, we find that a rotation of the pole-pieces through an angle ζ about the trajectory axis produces a perturbation of the variational function of the type (7.7.8), where

$$\eta = \pm 2 p a^{-2} n \zeta. \qquad (7.8.22)$$

If we may again assume that errors in different sectors are independent, the integrals (7.7.10) may be estimated as before. In this way we obtain from (7.7.20) the estimate

$$\epsilon_{\text{exp.}} \approx 3 N^{\frac{1}{2}} \zeta_{\text{r.m.s.}} \qquad (7.8.23)$$

for the expectation value of the half-width of the band situated at $\Theta_x + \Theta_y = 0$. If we again stipulate that the accelerator should be operating half a band width away from the edge of the band of instability even if $\Theta_x + \Theta_y$ approaches to within 3° of its critical value, the r.m.s. value of the angular misalinement must be less than 0·0013 radian, i.e. 4 minutes of arc.

7.9. Non-linear effects

We shall conclude this chapter with a brief investigation of the influence upon the stability of a periodic-focusing accelerator, such as the strong-focusing synchrotron, of non-linear terms in the trajectory equations. Non-linear effects have not been studied to anything like the same extent as linear perturbations, so that, while general formulae and conclusions may be advanced with some confidence, tolerances which are estimated in this section must, for reasons which are given later on, be considered tentative.

We saw in §§ 6.5 and 7.4 that it is possible to regard the effect of non-linear terms as a perturbation problem which may be treated by the method of variation of parameters. However, the only problem which we have so far considered explicitly, that of coupling between radial and phase oscillations in a linear accelerator, was such that the most interesting effect could be studied by making certain reasonable but restrictive assumptions which effectively reduced the problem to one on linear perturbations. It seems that, in general, non-linear terms produce instability at small amplitudes in a uniform-focusing accelerator (assumed free from periodic perturbations) only if it effects a coupling between normal modes. In a periodic-focusing accelerator, on the other hand, non-linear terms may produce instability, for small amplitudes, even if these terms do not effect coupling between different modes of oscillation; this will be demonstrated in this section by calculations which follow the pattern laid down in § 6.5. It is worth recalling that the results established in § 6.5 by perturbation theory were in agreement with the known analytical solution of the equation there considered, since support for the results of § 6.5 is to some extent support for the results of the present section also.

We shall begin by considering the term $M^{(3)}$ in the expansion (6.2.21) for the variational function, since, after the term $M^{(2)}$, this may be considered the most important. However, since we do not intend to consider coupling effects in this section, we shall deal only with that part of $M^{(3)}$ which involves one particular co-ordinate; this we take to be x.

We may form from $M^{(3)}$ a perturbation characteristic function, which we write as $U^{(3)}$, by means of (7.4.1). On noting that this function is a cubic polynomial in p_x and x, we see from (7.2.19) that the function may be expressed in the form

$$U^{(3)} = K^3 F(\kappa), \tag{7.9.1}$$

where F must be expressible as

$$F(\kappa) = A \sin(\kappa + \alpha) + B \sin(3\kappa + \beta). \tag{7.9.2}$$

The suffix k has been suppressed.

We may now obtain, with the help of (7.4.5) and (7.4.7), formulae for the change in amplitude and phase over one revolution of the accelerator due to cubic terms in the variational function. However, it is proposed that we absorb into $\Delta\kappa$ the phase change Θ due to the paraxial properties of the accelerator; thus we obtain

$$\Delta K = \sigma^{-1} K^2 F' \tag{7.9.3}$$

and

$$\Delta\kappa = \Theta - 3\sigma^{-1} K F, \tag{7.9.4}$$

where F' denotes the derivative of F with respect to κ.

Let us now consider the increment (7.9.3) summed over a large number of revolutions, assuming that the amplitude is sufficiently small for the overall change in K to be regarded as small. Then if, for the time being, we also suppose that the second term on the right-hand side of (7.9.4) is small compared with Θ, the overall change in K is the sum of two parts; these parts involve the summation of series of which the kth terms are $\cos(k\Theta+\alpha)$ and $\cos(3k\Theta+\beta)$. It is clear that if Θ is close to zero, both these series will oscillate with large amplitude so that one has the likelihood of instability in the machine. If Θ is close to $\pm\frac{2}{3}\pi$, the second series builds up to large amplitude although the first series does not. If Θ does not approximate to any of these critical values, our present calculations give no reason to expect the term under consideration to produce instability.‡

We shall now investigate the possibility of instability when Θ is close to zero by examining the equations (7.9.3) and (7.9.4) with the assumption that Θ is sufficiently small to be comparable with the second term on the right-hand side of (7.9.4). We may infer from the preceding paragraph that *the following calculations may be applied to the study of instability near to* $\Theta=\pm\frac{2}{3}\pi$ *by replacing* Θ *by* $\Theta-\frac{2}{3}\pi$ *or* $\Theta+\frac{2}{3}\pi$, *respectively, and omitting the first term on the right-hand side of* (7.9.2).

Apart from the assumptions entailed in first-order perturbation theory, the basic assumption of this section is that *the differences given by* (7.9.3) *and* (7.9.4) *are sufficiently small to be regarded as differentials.* By using this approximation, we may relate ΔF to $\Delta\kappa$ and so obtain from (7.9.3) and (7.9.4) the relation

$$\sigma\Theta K\Delta K-(K^3\Delta F+3FK^2\Delta K)=0, \qquad (7.9.5)$$

which may be integrated to give

$$\tfrac{1}{2}\sigma\Theta K^2-FK^3=C, \qquad (7.9.6)$$

where C is a constant. This equation determines the relation between the amplitude and phase of a trajectory in terms of its initial amplitude and phase.

The question to be answered by reference to (7.9.6) and to the fact that the function $F(\kappa)$ is bounded, is the following: What

‡ It is important to notice that even if F contains a term independent of κ, as will be the case when non-linear effects of even order are considered, this term disappears on differentiation and so does not appear in (7.9.3).

restriction must one place upon the amplitude K, or the phase shift Θ, in order that one may be able to assert that the amplitude will vary only slightly during all subsequent revolutions of the accelerator? In order to simplify this problem, let us assume that the upper and lower bounds of $F(\kappa)$ are F_M and $-F_M$; if these bounds are farther apart than is really necessary, our estimates will be over-cautious.

If we now make the substitutions

$$u = 2\sigma^{-1}\Theta^{-1}F_M K, \quad v = F/F_M, \tag{7.9.7}$$

and relate c to c accordingly, (7.9.6) may be rewritten as

$$v = u^{-1} - cu^{-3}. \tag{7.9.8}$$

Assuming for definiteness that Θ is positive, we wish to determine from (7.9.8) from what values of u it is possible for the 'configuration point' (that is, the point representing the amplitude and phase of the trajectory) to reach $u = \infty$ by travelling along a curve $c = \text{const.}$, subject to the restriction that $-1 \leqslant v \leqslant 1$ (see fig. 7.6).

If c is negative, the curve $v = v(u)$ decreases monotonically to zero so that all configuration points on this curve, for which $|v| \leqslant 1$, may reach infinity without violating this condition. If c is positive, however, the curve $v = v(u)$ rises from $-\infty$ at $u = 0$ to a maximum value $v = 2/(3\sqrt{(3c)})$ at $u = \sqrt{(3c)}$, after which it decreases monotonically to zero. It is clear that all configuration points (for which $|v| \leqslant 1$) to the right of the maximum may reach infinity, but that points to the left of the maximum may reach infinity only if the maximum value of v for that curve is not greater than unity. Hence all configuration points in the region of the u-v plane bounded by the v-axis, the lines $v = \pm 1$, and the curve $c = \frac{4}{27}$ are 'trapped' in the sense that they cannot escape to infinity by moving along a curve $c = \text{const.}$ without going outside the limits set by $|v| \leqslant 1$. The curve $c = \frac{4}{27}$ cuts the line $v = -1$ at $u = \frac{1}{3}$ and the line $v = 1$ at $u = \frac{2}{3}$. From this we deduce that any configuration point for which $u < \frac{1}{3}$ is trapped, any point for which $u \geqslant \frac{2}{3}$ is 'free' in the sense that it can reach infinity, but that a point for which $\frac{1}{3} \leqslant u < \frac{2}{3}$ may be either trapped or free depending upon its values of v.

If we note that the above rules may be retained with only slight alteration if Θ is negative, we see that these rules may be expressed in terms of our original variables as follows: Any trajectory for which

$K < \frac{1}{6}\sigma F_M^{-1} \mid \Theta \mid$ will remain bounded for all time, and any trajectory for which $K \geqslant \frac{1}{3}\sigma F_M^{-1} \mid \Theta \mid$ will expand indefinitely; but if

$$\tfrac{1}{6}\sigma F_M^{-1} \mid \Theta \mid \leqslant K < \tfrac{1}{3}\sigma F_M^{-1} \mid \Theta \mid,$$

the trajectory may either remain bounded or expand continuously, depending upon its phase.

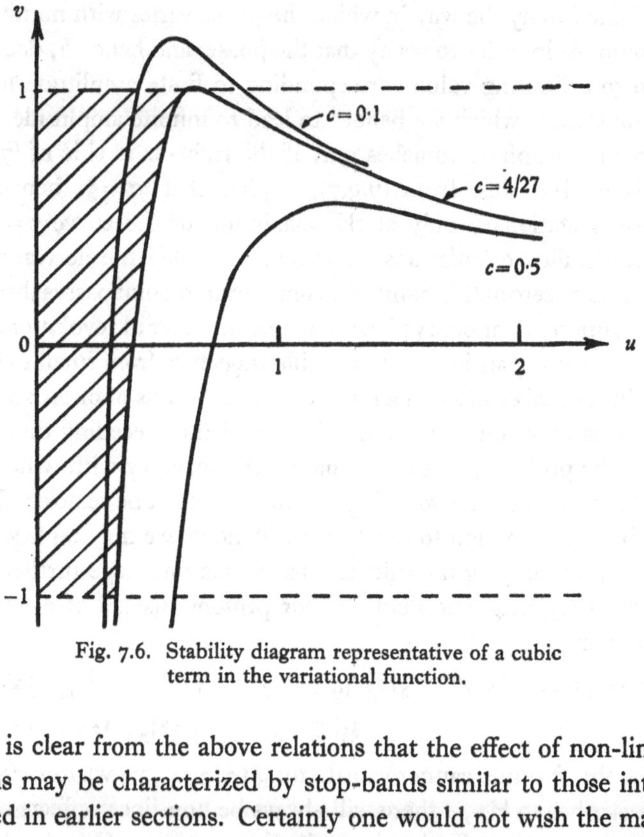

Fig. 7.6. Stability diagram representative of a cubic term in the variational function.

It is clear from the above relations that the effect of non-linear terms may be characterized by stop-bands similar to those introduced in earlier sections. Certainly one would not wish the maximum amplitude of the largest stable trajectory to be less than the half-aperture, w, of the accelerator. This suggests that we specify that $\mid \Theta \mid > \Theta_0$, where

$$\Theta_0 = 3\sigma^{-1} F_M w. \tag{7.9.9}$$

However, we have already seen that at the edge of this stop-band the largest stable orbit drifts in amplitude over a ratio of 2 : 1 so that the available aperture is being used uneconomically. It would be better to specify that the variation in amplitude should not exceed, say, $\pm 10\%$; it is found that this trebles the width of the stop-band given by (7.9.9).

Apart from the above stop-band, centred at $\Theta=0$, there will be two other stop-bands, of half-widths comparable with Θ_0, centred at $\Theta=\pm\frac{2}{3}\pi$.

Before we proceed to apply the above results, it is as well to investigate briefly the way in which the phase varies with number of revolutions in order to verify that the phase, and hence F, does not tend to a limiting value corresponding to finite amplitude under circumstances which we believe to lead to infinite amplitude. The increment in phase vanishes only if the right-hand side of (7.9.4) vanishes, but this, from (7.9.7), implies that $uv=\frac{2}{3}$; hence the phase is stationary only at the maximum of the curve $v=v(u)$. More detailed calculations show that, provided the derivative of $F(\kappa)$ is non-zero at this point, the configuration point passes through the position of stationary phase in a finite number of revolutions. In general, the expansion of an unstable trajectory from small to large amplitudes takes place in a number of revolutions of order $2\pi/\Theta$.

Let us now consider the application of the preceding theory to a specific problem, that of estimating the accuracy with which the field-form of a strong-focusing synchrotron must be realized. If we confine our attention to betatron oscillations, we may replace $M^{(3)}$ by $m^{(3)}$; by carrying the calculations of § 5.2 one stage further, but for a purely magnetic field and for protons instead of electrons, we obtain‡

$$m^{(3)}=\tfrac{1}{2}\mathsf{p}\kappa x(x'^2+y'^2)-\tfrac{1}{3}\mathsf{H}'_y y(x'y-xy')+\tfrac{1}{6}(2\kappa\mathsf{H}_{y,x}-\mathsf{H}_{y,xx})x^3$$
$$+\tfrac{1}{6}(\mathsf{H}''_y-3\kappa\mathsf{H}_{y,x}+3\mathsf{H}_{y,xx})xy^2. \quad (7.9.10)$$

Since this formula cannot vanish, for all x, y, x', y', whatever value one ascribes to $\mathsf{H}_{y,xx}$, there will always be non-linear effects of the type we are investigating in a circular machine. One would, of course, investigate the non-linear effects in a perfect accelerator by considering one section of the machine rather than a complete revolution; in this way we should discover forbidden values of θ. However, since it requires finer adjustment to avoid forbidden values of Θ, it is proposed that we examine stop-bands, due to non-

‡ This formula is a special case of equation (4.10) of ref. (37).

linear effects, which are peculiar to imperfect machines. Since we have already considered, in the linear approximation, variations in H_y and $H_{y,x}$, we shall consider here variations in $H_{y,xx}$. In this section we are also confining our attention to the x co-ordinate, so that we may for present purposes replace (7.9.10) by the single term

$$m^{(3)} = -\tfrac{1}{6}H_{y,xx}x^3. \qquad (7.9.11)$$

In order to evaluate $U^{(3)}$, it is necessary to integrate (7.9.11) over a revolution. However, we saw earlier in this section that we are interested only in the maximum value of $|F|$, and this we may obtain from the maximum value of $|U^{(3)}|$. Moreover, we may verify from fig. 7.3 that, if the system operates in the $\frac{1}{2}\pi$ mode with reference planes determined by (7.2.31), the average value of x^3 is slightly less than K^3. If we now assume that $H_{y,xx}$ (or, more accurately, the *variation* in $H_{y,xx}$) is constant in any one sector but varies randomly from sector to sector, we obtain the following estimate for F_M:

$$F_M = \tfrac{1}{6}(2N)^{\frac{1}{2}} L(H_{y,xx})_{\text{r.m.s.}}. \qquad (7.9.12)$$

If, for convenience, we measure the field variation by the fractional change in field strength at the limit of the aperture by writing

$$\epsilon H_y = \tfrac{1}{2}(H_{y,xx})_{\text{r.m.s.}}\, w^2, \qquad (7.9.13)$$

and if we use (7.2.37), (7.8.1), (7.8.3), (7.8.7) and (7.8.8), the relation (7.9.9) becomes

$$\Theta_0 \approx 20 N^{-\frac{3}{2}}(a/w)\epsilon. \qquad (7.9.14)$$

As a possible example, we might consider that a, the radius of the accelerator, is 2500 cm. and w, the half-width of the aperture, is 10 cm.; then if, as before, we take $N=40$, (7.9.14) reduces to $\Theta_0 \approx 20\epsilon$. Hence if we specify that $\Theta_0 = 0{\cdot}05$, so that the width of the stop-band is about 6°, the maximum permissible fractional change in field-strength, *of the type we are considering*, is about $\frac{1}{4}$ %. In order to assess how stringent this requirement is, it would be necessary to have detailed information about the fluctuations in field-strength which are liable to arise in practice.

The method which has been applied to study non-linear effects in this section may be used to study other terms of the variational function including those which represent coupling between co-ordinates. One may in each case establish the existence of an invariant of the form (7.9.6), the second term of which is identical with (minus) the perturbation characteristic function associated with the term under consideration, and the first term of which is (minus) the perturbation characteristic function associated with a small phase shift Θ. This suggests that the following theorem is true: *the perturbation characteristic function which represents the (small) departure of a dynamical transformation from the identity transformation is an invariant of this transformation.* It is easy to verify that this theorem is true.

Let us suppose that the dynamical transformation, which we may take to represent one revolution (or a small number of revolutions) of an accelerator, is written as

$$p_r \to p_r + \Delta p_r, \quad x_r \to x_r + \Delta x_r, \tag{7.9.15}$$

where r enumerates the variables x_1, x_2, τ.

Then if the appropriate perturbation characteristic function is $\Delta U(p_r, x_r)$,

$$\Delta p_r = \frac{\partial \Delta U}{\partial x_r}, \quad \Delta x_r = -\frac{\partial \Delta U}{\partial p_r}. \tag{7.9.16}$$

Now since we are assuming Δp_r and Δx_r to be small, we may write the change in the function ΔU due to the change in its arguments as

$$\Delta(\Delta U) = \Delta p_r \frac{\partial \Delta U}{\partial p_r} + \Delta x_r \frac{\partial \Delta U}{\partial x_r}, \tag{7.9.17}$$

but substitution from (7.9.16) shows immediately that

$$\Delta(\Delta U) = 0, \tag{7.9.18}$$

so that ΔU is indeed an invariant.

The above invariant may be used to study other non-linear terms, and it may also be used to study linear perturbations such as those which were analysed in § 7.5 by matrix methods. Nevertheless, it was instructive to derive the invariant (7.9.6) in detail, particularly as it proves advantageous to carry out a detailed derivation in analysing terms representing non-linear coupling.

We saw in §§ 7.3 and 7.7 that an invariant can provide a simple explanation of a result obtained by more complicated methods, and we have now seen in this section that an invariant can provide the solution to an interesting particular case of a difficult problem. These observations suggest that until there is available a comprehensive study of the invariants of periodic-focusing systems, our understanding of strong-focusing accelerators will be incomplete.

BIBLIOGRAPHY

The following is a selection of modern books on electron optics:

(1) V. K. ZWORYKIN, G. A. MORTON, E. G. RAMBERG, J. HILLIER and A. W. VANCE. *Electron Optics and the Electron Microscope* (Wiley, New York, 1945).

(2) J. R. PIERCE. *Theory and Design of Electron Beams* (van Nostrand, New York, 1949).

(3) V. E. COSSLETT. *Introduction to Electron Optics* (Oxford University Press, 2nd ed. 1950).

(4) W. GLASER. *Grundlagen der Elektronenoptik* (Springer, Vienna, 1952).

(5) O. KLEMPERER. *Electron Optics* (Cambridge University Press, 2nd ed. 1953).

The following deal with particle accelerators:

(6) *The Acceleration of Particles to High Energies* (Institute of Physics, London, 1950).

(7) J. C. SLATER. *Microwave Electronics* (van Nostrand, New York, 1950).

(8) M. S. LIVINGSTON. *High-energy Accelerators* (Interscience, New York, 1954).

Review articles, with extensive and up-to-date bibliographies frequently appear in

(9) *Advances in Electronics*, editor L. Marton (Academic Press, New York, approximately annually from 1948).

Reference may also be made to the following books on geometrical optics:

(10) J. L. SYNGE. *Geometrical Optics: An Introduction to Hamilton's Method* (Cambridge University Press, 1937).

(11) R. K. LUNEBERG. *Mathematical Theory of Optics* (Brown University, Providence, 1944).

And to the following book on classical dynamics:

(12) H. C. CORBEN and P. STEHLE. *Classical Mechanics* (Wiley, New York, 1950).

The following papers, grouped by chapters, are suggested for further reading:

Chapter 1

(13) W. GLASER. Über geometrisch-optische Abbildung durch Elektronenstrahlen. *Z. Phys.* **80** (1933), 451–64.

(14) F. BORGNIS. Allgemeine Eigenschaften der paraxialen elektronenoptischen Abbildung. *Helv. Phys. Acta*, **21** (1948), 461–79.

(15) P. GRIVET. Electron lenses. *Advanc. Electron.* **2** (1950), 47–100.
(16) G. LIEBMANN. Field plotting and ray tracing in electron optics. A survey of numerical methods. *Advanc. Electron.* **2** (1950), 101–149.

Chapter 2

(17) D. GABOR. Electron optics. *Elect. Engng*, **15** (1942–3), 295–9, 328–31 and 372–4.
(18) D. GABOR. Dynamics of electron beams. Applications of Hamilton's dynamics to electron problems. *Proc. Inst. Radio Engrs, N.Y.*, **33** (1945), 792–805.
(19) L. A. MACCOLL. The fundamental equations of electron motion (dynamics of high-speed particles). *Bell Syst. Tech. J.* **22** (1943), 153–77.
(20) F. GRAY. Electrostatic electron optics. *Bell Syst. Tech. J.* **18** (1939), 1–31.
(21) K. SPANGENBERG. Use of the action function to obtain the general differential equations of space-charge flow in more than one dimension. *J. Franklin Inst.* **232** (1941), 365–71.

Chapter 3

(22) L. A. MACCOLL. Trajectories of monoenergetic electrons, in an arbitrary static electromagnetic field, in the neighbourhood of a given trajectory. *J. Math. Phys.* **20** (1941), 355–69.
(23) R. G. E. HUTTER. Class of electron lenses which satisfy Newton's relation. *J. Appl. Phys.* **16** (1945), 670–8.
(24) W. GLASER and O. BERGMANN. Über die Tragweite der Begriffe 'Brennpunkte' und 'Brennweite' in der Elektronenoptik und die starken Elektronenlinsen mit Newtonscher Abbildungsgleichung. *Z. angew. Math. Phys.* **1** (1950), 363–79 and **2** (1951), 159–88.
(25) P. A. STURROCK. Perturbation characteristic functions and their application to electron optics. *Proc. Roy. Soc.* A, **210** (1951), 269–89.

Chapter 4

(26) W. GLASER. Zur Bildfehler des Elektronenmikroskops. *Z. Phys.* **97** (1935), 177–201.
(27) O. SCHERZER. Über einige Fehler von Elektronenlinsen. *Z. Phys.* **101** (1936), 593–603.
(28) O. SCHERZER. Sphärische und chromatische Korrektur von Elektronenlinsen. *Optik*, **2** (1947), 114–32.
(29) W. GLASER. Strenge Berechnung magnetischen Linsen der Feldform $H = H_0/(1+(z/a)^2)$. *Z. Phys.* **117** (1941), 285–315.
(30) W. GLASER and E. LAMMEL. Strenge Berechnung der elektronenoptischer Aberrationskurven eines typischen Magnetfelder. *Arch. Elektrotech.* **37** (1943), 347–56.
(31) R. G. E. HUTTER. Rigorous treatment of the electrostatic immersion lens whose axial potential distribution is given by $\phi(z) = \phi_0 \exp(K \arctan z)$. *J. Appl. Phys.* **16** (1945), 678–99.

(32) E. Regenstreif. Théorie de la lentille électrostatique à trois électrodes. *Ann. Radioelect.* **6** (1951), 51–83 and 114–55.

(33) W. Glaser. Zentrierung und Auflösungsvermogen beim Übermikroskop. *Öst. Ingen.-Arch.* **3** (1949), 39–46.

(34) F. Bertein. Quelque défauts des instruments d'optique électronique et leur correction. *Ann. Radiolect.* **2** (1947), 379–408 and **3** (1948), 49–62.

(35) P. A. Sturrock. The aberrations of magnetic electron lenses due to asymmetries. *Phil. Trans.* A, **243** (1951), 387–429.

Chapter 5

(36) M. Cotte. Recherches sur l'optique électronique. *Ann. Phys., Paris*, **10** (1938), 333–405.

(37) P. A. Sturrock. The imaging properties of electron beams in arbitrary static electromagnetic fields. *Phil. Trans.* A, **245** (1952), 155–87.

(38) O. Persico and C. Geoffrin. Beta-ray spectroscopes. *Rev. Sci. Instrum.* **21** (1950), 945–70.

(39) P. Grivet. Les spectrographes β à lentilles électroniques. (Théorie unifiée des types classiques: un nouvel appareil.) *J. Phys. Radium*, **11** (1950), 582–95 and **12** (1951), 1–14.

(40) H. von Marschall. Grundlegung der elektronenoptischen Theorie eines Massenspektrographen. *Phys. Z.* **45** (1944), 1–37.

(41) M. G. Inghram. Modern mass spectroscopy. *Advanc. Electron.* **1** (1948), 219–68.

(42) J. Picht and J. Himpan. Beiträge zür Theorie der elektrischen Ablenkung von Elektronenstrahlbündeln. *Ann. Phys., Lpz.*, **39** (1941), 409–501.

(43) W. Glaser. Über die Theorie der elektrischen und magnetischen Ablenkung von Elektronenstrahlen und ein ihr angepasstes Störungverfahren. *Ann. Phys., Lpz.*, **4** (1949), 389–408.

(44) R. G. E. Hutter. The deflection of beams of charged particles. *Advanc. Electron.* **1** (1948), 167–218.

Chapter 6

(45) D. Bohm and L. Foldy. The theory of the synchrotron. *Phys. Rev.* **70** (1946), 249–58.

(46) J. S. Gooden, H. H. Jensen and J. L. Symonds. Theory of the proton synchrotron. *Proc. Phys. Soc.* B, **59** (1947), 677–93.

(47) R. Q. Twiss and N. H. Frank. Orbital stability in a proton synchrotron. *Rev. Sci. Instrum.* **20** (1949), 1–17.

(48) D. W. Fry and W. Walkinshaw. Linear accelerators. *Rep. Progr. Phys.* **12** (1949), 102–32.

(49) J. R. Terral and J. C. Slater. Particle dynamics in the linear accelerator. *J. Appl. Phys.* **23** (1952), 66–77.

Chapter 7

(50) N. M. Blachman and E. D. Courant. The dynamics of a synchrotron with straight sections. *Rev. Sci. Instrum.* **20** (1949), 596–601.

(51) N. M. BLACHMAN. Forced betatron oscillation in a synchrotron with straight sections. *Rev. Sci. Instrum.* **21** (1951), 569–571.

(52) C. HENDERSON, F. F. HEYMANN and R. E. JENNINGS. Phase stability of the microtron. *Proc. Phys. Soc.* B, **66** (1953), 41–9.

(53) J. S. BELL. Vertical focusing in the microtron. *Proc. Phys. Soc.* B, **66** (1953), 802–4.

(54) K. J. LECOUTEUR. The regenerative deflector for synchrocyclotrons. *Proc. Phys. Soc.* B, **64** (1951), 1073–84.

(55) E. D. COURANT, M. S. LIVINGSTON and H. S. SNYDER. The strong-focusing synchrotron—a new high-energy accelerator. *Phys. Rev.* **88** (1952), 1190–6.

(56) J. P. BLEWETT. Radial focusing in the linear accelerator. *Phys. Rev.* **88** (1952), 1197–8.

(57) J. B. ADAMS, M. G. N. HINE and J. D. LAWSON. Effect of magnetic inhomogeneities in the strong-focusing synchrotron. *Nature, Lond.*, **171** (1953), 926–7.

(58) A. M. CLOGSTON and H. HEFFNER. Focusing of an electron beam by periodic fields. *J. Appl. Phys.* **25** (1954), 436–47.

INDEX

Printed in the United States
By Bookmasters